Springer Tracts in Modern Physics 106

Editor: G. Höhler
Associate Editor: E. A. Niekisch

Editorial Board: S. Flügge H. Haken J. Hamilton
H. Lehmann W. Paul

Springer Tracts in Modern Physics

* denotes a volume which contains a Classified Index starting from Volume 36

J. Kirschner

Polarized Electrons at Surfaces

With 57 Figures

Springer-Verlag Berlin Heidelberg GmbH

Dr. Jürgen Kirschner

Institut für Grenzflächenforschung und Vakuumphysik der Kernforschungsanlage,
Postfach 1913, D-5170 Jülich 1, Fed. Rep. of Germany

Manuscripts for publication should be addressed to:

Gerhard Höhler

Institut für Theoretische Kernphysik der Universität Karlsruhe
Postfach 6380, D-7500 Karlsruhe 1, Fed. Rep. of Germany

*Proofs and all correspondence concerning papers in the process of publication
should be addressed to:*

Ernst A. Niekisch

Haubourdinstrasse 6, D-5170 Jülich 1, Fed. Rep. of Germany

Library of Congress Cataloging in Publication Data. Kirschner, Jürgen, 1945– Polarized electrons at surfaces.
(Springer tracts in modern physics; 106)) Bibliography: p. Includes index. 1. Electrons – Polarization. 2. Surfaces
(Physics) I. Title. II. Series. QC1.S797   vol. 106   539 s   [530.4'1]   84-26836   [QC793.5.E628]

This work is subject to copyright. All rights are reserved, whether the whole or part of the material is concerned,
specifically those of translation, reprinting, reuse of illustrations, broadcasting, reproduction by photocopying
machine or similar means, and storage in data banks. Under § 54 of the German Copyright Law where copies are
made for other than private use, a fee is payable to "Verwertungsgesellschaft Wort", Munich.

ISBN 978-3-662-15219-5   ISBN 978-3-540-39212-5 (eBook)
DOI 10.1007/978-3-540-39212-5

© Springer-Verlag Berlin Heidelberg 1985

Originally published by Springer-Verlag Berlin Heidelberg New York in 1985.
Softcover reprint of the hardcover 1st edition 1985

The use of registered names, trademarks, etc. in this publication does not imply, even in the absence of a specific
statement, that such names are exempt from the relevant protective laws and regulations and therefore free for
general use.

2153/3130-543210

Preface

This book covers a section of the field of polarized electron physics, which has experienced a particularly rapid growth over the past decade. It deals with spin-polarized electron spectroscopies applied to solid surfaces, with particular emphasis on the information that can be obtained by observing the spin of scattered or emitted electrons. Both magnetic and non-magnetic solids are investigated. Depending on the type of experiment and the range of electron energies involved mainly surface properties or mainly bulk properties are probed, though a sharp distinction cannot always be made.

The book addresses itself to researchers in the field, to graduate physics students, and to all those who would like to know more about this new branch of surface physics. The emphasis is on basic concepts for the description of polarized electrons at surfaces and on recent experimental results and their interpretation.

The present work is largely based on the "Habilitationsschrift" of the author, submitted to the Rheinisch-Westfälische Technische Hochschule Aachen.

I gratefully acknowledge the invaluable support by Prof. H. Ibach, many helpful discussions with colleagues, the excellent work of Mrs. Krüger and Mr. Larscheid, and, above all, the patience and understanding of my wife Gudrun.

Jülich, March 1985   *J. Kirschner*

Contents

1. Introduction

Spin is a genuine property of the electron. Although it was postulated as early as 1925 by Goudsmit and Uhlenbeck /1925/ we still do not 'really know' what it is, in the sense that we could associate some phenomenon out of our everyday experience with the electron spin. We do know that it is a measurable quantum mechanical variable which obeys the angular momentum formalism, that there is a magnetic moment associated with it, and that its half-integer value puts the electron into the class of fermions, for which the Pauli principle is valid. On the other hand, we know that it is not necessary to have a classical analogue to a quantum mechanical property in order to understand a physical phenomenon. For example, we learned to live quite comfortably with the wave-particle dualism of the electron. We learned to exploit the information contained in its energy and momentum in a number of surface spectroscopies that were developed during the past three decades. The electron spin as a source of information about surface and bulk properties of solids has been harnessed only since about a decade ago. The present work is intended to give an introduction to this field and to describe its present status.

There are two main physical phenomena giving rise to spin-dependent effects in the interaction of free electrons with surfaces: the spin-orbit interaction and the exchange interaction. Both effects in general are present simultaneously, interfering with each other, but there are conditions where they may be isolated to some extent and studied separately. The spin-orbit interaction is most pronounced in electron scattering or emission from heavy, non-magnetic materials. It is essentially due to the interaction of the spin of an electron with its own angular momentum in the electric field of a strong scattering potential. The important ingredient therefore is the presence of a strongly attractive ion core. The conduction electrons and bound electrons do play a role, but they are not essential for the existence of the effect. The exchange interaction, however, is bound to the presence of other electrons. It is essentially an outcome of the Pauli principle, requiring the total wavefunction (including spins) of fermions to be antisymmetric with respect to a permutation of the particles. The forces arising from the exchange interaction are of Coulombic nature, and are therefore comparable to those due to spin-orbit interaction. When scattering an electron from a surface, exchange interactions are always involved, even with paramagnetic materials. However, as spin-up and spin-down electrons are present in exactly equal amounts in these materials, there is no net effect on the spin polarization or relative intensities of scattered or emitted electrons due to the interaction with the electrons of the solid. In ferromagnets, this balance is destroyed and we have polarization effects even if spin-orbit interaction is negligible. This allows us to study surface and bulk magnetism and the electronic structure of ferromagnets via an analysis of the spin state of scattered or emitted electrons.

Like a number of modern developments in surface physics, the field of surface physics with spin-polarized electrons dates back to experiments in the nineteen-twenties 'the Golden Age of physics', which at that time had no or unsatisfactory results. As is well-known, Davisson and Germer in 1927 demonstrated the wave nature of the electron by diffraction from the surface of a Ni single crystal /Davisson and Germer, 1927/. It is less well-known that shortly afterwards they started searching for the polarization properties of these 'electron waves'. They conceived a mirror experiment for electrons, in close analogy to a double mirror experiment for visible light. Their schematic is reproduced in Fig. 1.1.

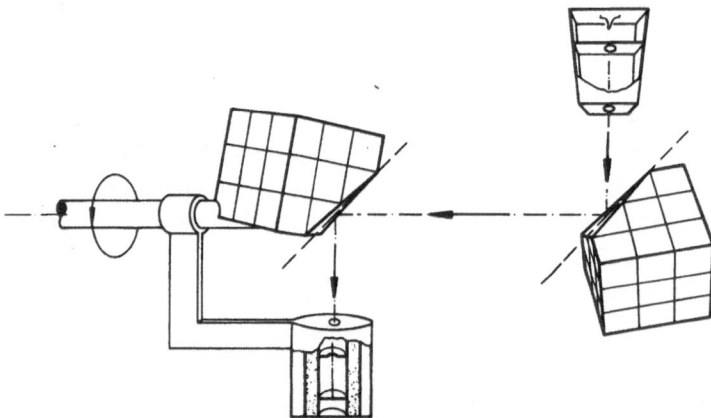

Fig. 1.1. Schematic view of the early experiment by Davisson and Germer /1929/ aimed at "polarizing electron waves". On the right hand side unpolarized primary electrons from a hair-pin cathode impinge onto the (111) surface of a Ni single crystal ('polarizer'). The elastically reflected electrons are sent to a second Ni(111) crystal ('analyzer') at the same angle of incidence. Rotation of the analyzer assembly about the horizontal axis was expected to result in intensity fluctuations of the current measured in the Faraday cup collector

Unpolarized electron waves are polarized by reflection from the first Ni(111) surface, in analogy to the linear polarization of light by reflection at the Brewster angle. This 'polarizer' is followed by an identical 'analyzer', i.e. a further reflection from a Ni(111) surface. Turning the analyzer about the axis of the incident light should give rise to characteristic intensity variations of the reflected electrons. If the electron waves were linearly polarized, one would expect a periodicity with $n\pi$ (n = 1, 2, ...), as is observed with light. This early experiment did not give any effect within an error margin of 0.5 % and Davisson and Germer /1929/ concluded 'electron waves are not polarized by reflection'. Similar attempts by a number of other workers did not lead to any other result (see the extended historical work by Gehrenbeck /1973/). Soon afterwards, Weisskopf /1935/ gave arguments, why sizeable polarization effects should not be expected. As a consequence of the negative results of the experiments and their theoretical confirmation there were no further attempts to show spin-dependent effects in electron diffraction from solids. For atoms, Mott /1932/ had developed the theory of electron-atom scattering on the basis of Dirac's /1928/ theory of the electron, and predicted polarization effects due to spin-orbit coupling. A number of corresponding atomic physics experiments were carried out over the following decade, but their results turned out to be ambiguous or contradictory. Not earlier than 1943 were Shull et al. /1943/ able to measure definite polarization effects. The first experimental evidence for polarization effects in scattering from solids was obtained by Loth /1967/ and Eckstein /1967/ for W, Au and Pt. They used polycrystalline, uncharacterized and probably contaminated surfaces. The results showed some similarity to those from free atoms, with a strongly reduced degree of polarization. These, as well as a number of other successful atomic physics experiments (see the review by Kessler /1976/), renewed the interest in spin-polarized electron diffraction at the beginning of the seventies. Maison /1966/ pointed out errors in the theoretical treatment by Weisskopf who had expanded the lattice potential into a Fourier series and cut off after a few terms. The higher terms, however, describe the potential gradients near the ion cores, which in turn determine the strength of the spin-orbit interaction. Thus the early result of Weisskopf was due to the neglect of the higher Fourier coefficients. With respect to the early experiment of Davisson and Germer it was pointed out by Kuyatt /1975/ that they had looked for the wrong periodicity. The polarization vector has the transformation properties of an angular momentum, and is thus an axial vector or pseudovector. In contrast to the result for a polar vector (e.g. linearly polarized light), for which a periodicity with $n\pi$ is found, for an axial vector a periodicity of $2n\pi$ is to be expected.

The quantitative theoretical treatment of polarized electron diffraction started with the work of Jennings /1970, 1971a, 1971b/ and Feder /1971, 1972/, while the first experimental demonstration of the polarization of electrons by reflection was achieved by O'Neill et al. /1975/ on W(001). The first sucessful realization of a double scattering experiment with two crystal surfaces was reported 50 years after its first attempt, by Kirschner and Feder /1979/. They also observed intensity asymmetries in diffracted, normally equivalent beams, when using a polarized primary beam, which led to the development of a new type of polarized electron detector. While the above experiments investigated spin-orbit interaction, the usefulness of polarized electrons in the study of exchange interactions was demonstrated first by Celotta et al. /1979/ for ferromagnetic Ni. In electron emission from ferromagnets the existence of sizeable spin polarization was noted relatively early. Pioneer experiments were carried out in photoemission by Busch et al. /1969/, in field emission by Müller et al. /1972/, and in secondary electron emission by Chrobok and Hofmann /1976/.

The present book treats non-magnetic and magnetic materials separately, according to the two main spin-dependent interactions, though not without mentioning possible interferences. It is not intended to serve only the specialist, and therefore contains an extended chapter (Chap. 2) introducing to the description of polarized electrons, and the spin-orbit and the exchange interactions. A central topic when dealing with electrons at surfaces is the 'LEED state', not only for electron scattering but for electron emission as well. A short survey of photo-emission and inelastic scattering concludes Chap. 2. In Chap. 3 we consider more recent experimental developments, and various types of sources and detectors for polarized electrons. Results from non-magnetic materials are discussed in Chap. 4, where we distinguish between electron diffraction and photoelectron emission. The latter topic starts with momentum- and spin-resolved photoemission, as the field of photo-yield spectroscopies is well covered by other review articles (Alvarado et al. /1978/, Meier and Pescia /1984/, Siegmann et al. /1984/, Kessler /1976/). In Chap. 5 results from magnetic surfaces are discussed: elastic and inelastic scattering from crystalline and amorphous surfaces, and various emission spectroscopies. An outlook to future developments closes the book. The list of references is not meant to be exhaustive, but the aim was to include at least a few relevant papers for each topic where the reader interested in more details may find additional references. In addition to the review papers already mentioned, the articles by Celotta and Pierce /1980/, Feder /1981/, and Pierce and Celotta /1982/ are recommended for further information.

2. Basic Concepts

This chapter contains the basic notions and concepts used in the subsequent chapters. After a short recapitulation of the formalism used for the description of ensembles of polarized electrons (Sect. 2.1), the two main types of spin-dependent interactions are discussed: Spin-orbit interaction (Sect. 2.2) and exchange interaction (Sect. 2.3). Spin-orbit interaction is treated within the context of electron scattering from a central potential. Exchange interaction is demonstrated to arise from the Pauli principle in the interaction of two spin-1/2 particles. Section 2.4 describes the concepts developed to calculate spin-dependent elastic diffraction from non-magnetic and magnetic surfaces. The spin-dependent LEED state defined there reappears in Sect. 2.5 in the context of spin-polarized photoemission. Using the information contained in the electron spin is shown to considerably enlarge the knowledge of the electronic band structure of magnetic and non-magnetic solids. A brief discussion of spin-dependent inelastic processes closes the chapter. The emphasis is on a description of the physical processes and of the assumptions underlying the theoretical treatment. For more detailed discussions reference is made to several excellent review papers or books.

2.1 Description of Polarized Electrons

The quantum mechanical property 'spin' is represented by an operator

$$\underline{s} = (s_x, s_y, s_z)$$

which satisfies the commutation rules

$$s_x s_y - s_y s_x = -i\hbar s_z \qquad \text{and cyclic.} \tag{2.1}$$

Introducing the Pauli spin matrices $\underline{\sigma} = (\sigma_x, \sigma_y, \sigma_z)$ the spin operator may be written

$$\underline{s} = \frac{\hbar}{2}\underline{\sigma} \quad \text{with } \sigma_x = \begin{pmatrix} 0 & 1 \\ 1 & 0 \end{pmatrix}; \ \sigma_y = \begin{pmatrix} 0 & -i \\ i & 0 \end{pmatrix}; \ \sigma_z = \begin{pmatrix} 1 & 0 \\ 0 & -1 \end{pmatrix} \ . \tag{2.2}$$

The matrices σ_i are unitary and self-adjunct.

General spin states, for example a beam of electrons along the z axis with partial or complete alignment of the projection of the spin along the z axis can be
characterized by the spin function χ with complex coefficients a_1 and a_2:

$$\chi = a_1|\alpha> + a_2|\beta> = a_1\binom{1}{0} + a_2\binom{0}{1} = \binom{a_1}{a_2} \quad . \tag{2.3}$$

Without loss in generality the basis functions $|\alpha> = \binom{1}{0}$ and $|\beta> = \binom{0}{1}$ were used,
which are eigenfunctions of σ_z with eigenvalues ± 1. According to the statistical
interpretation of quantum mechanical wavefunctions the squares $|a_1|^2$ and $|a_2|^2$
give the probability to find the value $+\hbar/2$ or $-\hbar/2$ with respect to the z axis in
a measurement. More precisely, this means that the spin vector of length $s = \hbar\sqrt{3/4}$
lies somewhere on a cone around the z axis, so that its projection onto the axis
has the value $\hbar/2$ or $-\hbar/2$.

The polarization $\underline{P}$ is a vector quantity defined by the expectation value of the
spin operator $\underline{\sigma}$:

$$\underline{P}: = \frac{<\chi|\underline{\sigma}|\chi>}{<\chi|\chi>} \quad . \tag{2.4}$$

The degree of polarization P_z with respect to a given axis z is a scalar of magnitude $-1 \leq P_z \leq 1$ given by:

$$P_z = \frac{<\chi|\sigma_z|\chi>}{<\chi|\chi>} = \frac{|a_1|^2 - |a_2|^2}{|a_1|^2 + |a_2|^2} \quad . \tag{2.5}$$

Therefore, the degree of polarization P_z of a beam of electrons polarized along z
is:

$$P_z = \frac{N^\uparrow - N^\downarrow}{N^\uparrow + N^\downarrow} \tag{2.6}$$

where $N^\uparrow$ ($N^\downarrow$) is the number of electrons with spin parallel (antiparallel) to the
z axis.

More precisely, this is only valid in the rest frame of the electrons, as the
above definition is not Lorentz invariant. At relativistic electron energies the
transversal component of the polarization vector in the laboratory system is
smaller by the factor $mc^2/(mc^2+E)$. However, this difference is neglected in the
following as for all kinetic energies E encountered here $E \ll mc^2 = 0.51$ MeV
/Kessler, 1976/.

A partially polarized beam can be represented by a mixture of n pure spin
states with the polarization $\underline{P}(n)$:

$$\underline{p}^{(n)}: = \frac{<\chi^{(n)}|\underline{\sigma}|\chi^{(n)}>}{<\chi^{(n)}|\chi^{(n)}>} \quad . \tag{2.7}$$

The total polarization then is

$$\underline{P} = \frac{\sum\limits_n \langle \chi^{(n)} | \underline{\sigma} | \chi^{(n)} \rangle}{\sum\limits_n \langle \chi^{(n)} | \chi^{(n)} \rangle} \quad .$$ (2.8)

Defining the spin density matrix $\rho^{(n)}$ for the pure spin state (n) by:

$$\rho^{(n)} := \begin{pmatrix} a_1^{(n)} \\ a_2^{(n)} \end{pmatrix} \cdot (a_1^{(n)*}, a_2^{(n)*}) = \chi\chi^\dagger$$ (2.9)

the total spin density matrix is given by the sum of the n density matrices:

$$\rho = \sum_n \rho^{(n)} = \sum_n \begin{pmatrix} |a_1^{(n)}|^2 & a_1^{(n)} a_2^{(n)*} \\ a_1^{(n)*} a_2^{(n)} & |a_2^{(n)}|^2 \end{pmatrix} \quad .$$ (2.10)

As by definition $\sum\limits_n \langle \chi^{(n)} | \chi^{(n)} \rangle = \sum\limits_n (|a_1^{(n)}|^2 + |a_2^{(n)}|^2) = \mathrm{tr}\,\rho$ the polarization

vector $\underline{P}$ can be written

$$\underline{P} = \frac{\mathrm{tr}(\rho\underline{\sigma})}{\mathrm{tr}\,\rho} \quad \text{or} \quad P_i = \frac{\mathrm{tr}(\rho\sigma_i)}{\mathrm{tr}\,\rho} \quad i = x,y,z \quad .$$ (2.11)

Conversely, the spin density matrix can be expressed in the components of the polarization vector:

$$\frac{\rho}{\mathrm{tr}\,\rho} = \frac{1}{2} (\underline{1} + \underline{P}\,\underline{\sigma}) \quad .$$ (2.12)

The spin density matrix can be diagonalized by a transformation of the coordinate system, such that the z axis points into the direction of the polarization vector. After that it assumes the simple form

$$\rho' = \frac{1}{2} \begin{pmatrix} 1+P & 0 \\ 0 & 1-P \end{pmatrix} \quad .$$ (2.13)

By analogy with the scattering matrix in non-relativistic scattering theory a (2x2) spin-scattering matrix $\underline{S}$ may be introduced. The spin-scattering matrix relates the spin function χ of an ensemble of electrons before scattering to that after scattering χ':

$$\chi' = \underline{S}\,\chi \quad .$$ (2.14)

The scattering matrix is useful for the theoretical treatment of the scattering process. In particular, if certain time or space symmetries are given by the experiment, an analysis of the symmetry properties of the scattering matrix yields general information about certain components or invariance properties of the polarization (see Chap. 4).

The spin density matrix ρ' for the state after scattering is expressed by means of the scattering matrix:

7

$$\rho' = \chi'\chi'^\dagger = \underline{S} \chi (\underline{S} \chi)^\dagger = \underline{S} \chi\chi^\dagger \underline{S}^\dagger = \underline{S} \rho \underline{S}^\dagger \qquad (2.15)$$

as $\chi\chi^\dagger = \rho$, the spin density matrix of the initial state.

Using (2.12), the final state spin-density matrix ρ' can be expressed by means of the polarization $\underline{P}$ of the initial state and the spin-scattering matrix $\underline{S}$:

$$\rho' = \frac{1}{2} \underline{S} (\underline{1} + \underline{P} \underline{\sigma}) \underline{S}^\dagger \, \mathrm{tr}\, \rho \ . \qquad (2.16)$$

Now, using (2.11), which connects polarization $\underline{P}$ and spin-density matrix ρ, the polarization vector $\underline{P}'$ after scattering may be determined:

$$\underline{P}' = \frac{\mathrm{tr}(\rho'\underline{\sigma})}{\mathrm{tr}\, \rho'} = \frac{\mathrm{tr}(\underline{S}(\underline{1} + \underline{P} \underline{\sigma}) \underline{S}^\dagger \underline{\sigma})}{\mathrm{tr}(\underline{S}(\underline{1} + \underline{P} \underline{\sigma})\underline{S}^\dagger)} \ . \qquad (2.17)$$

An important special case is that of an unpolarized primary beam $\underline{P} = 0$. It follows

$$\underline{P}' = \frac{\mathrm{tr}\, (\underline{S} \underline{S}^\dagger \underline{\sigma})}{\mathrm{tr}\, (\underline{S} \underline{S}^\dagger)} \ . \qquad (2.18)$$

2.2 Spin-Orbit Interaction

In this section the spin-orbit coupling in elastic electron-atom scattering is treated. The effects arising from a periodic array of atoms are discussed in Sect. 2.4.

Let us consider the behaviour of an electron in an electromagnetic field. In general, the Dirac equation has to be used /Kessler, 1976/

$$\left[i\hbar \frac{\partial}{\partial t} -e\Phi + \underline{\alpha}(i\hbar c \nabla + e\underline{A}) - \beta mc^2 \right] \psi = 0 \qquad (2.19)$$

with the vector potential $\underline{A}$, the scalar potential Φ and $\underline{\alpha} = (\alpha_x, \alpha_y, \alpha_z)$ and β being (4x4) matrices. Accordingly, the wavefunction is a 4-component spinor. Even when dealing with low energy electrons the Dirac equation has to be used in electron atom scattering, as even slow electrons may become 'relativistic' in the vicinity of the nucleus.

However, the physical content is not well visible in the Dirac equation and we therefore will discuss the non-relativistic limit, which is given for both kinetic and potential energy small compared to mc^2:

$$\{ \underbrace{\frac{1}{2m} (\underline{p} - \frac{e}{c}\underline{A})^2 + e\Phi}_{I} \ - \ \underbrace{\frac{e\hbar}{2mc} \underline{\sigma}\, \mathrm{curl}\, \underline{A}}_{II} \ + \qquad (2.20)$$

$$+ \ \underbrace{i \frac{e\hbar}{4m^2 c^2} \underline{E}\, \underline{p}}_{III} \ - \ \underbrace{\frac{e\hbar}{4m^2 c^2} \underline{\sigma}\, [\underline{E} \times \underline{p}]\}}_{IV}\, \Psi \ = \ E\, \Psi \ .$$

Here $E + mc^2$ is the total energy. The two 'small' components of the four-spinor can be neglected in this limit. Thus the wavefunction Ψ is a two-component spinor.

Term I in (2.20) corresponds to the Hamilton operator of the non-relativistic Schrödinger equation for a 'spin-less' particle in an electromagnetic field. Term II describes the interaction of the spin with the magnetic field. When taking terms I and II together, we end up with the Pauli equation for a non-relativistic electron with spin. The third term is a relativistic correction to the energy and has no classical analogue. Term IV finally describes the spin-orbit interaction. For a central potential $\Phi = \Phi(r)$ the electric field $\underline{E}$ is given by $\underline{E} = -(1/e)d\Phi/dr \cdot \underline{r}/r$. With the definition of the angular momentum $\underline{l} = \underline{r} \times \underline{p}$ we obtain from IV the interaction energy U_{so} due to spin-orbit coupling:

$$U_{so} = -\frac{e\hbar}{4m^2c^2}\,\underline{\sigma}[\underline{E} \times \underline{p}] = \frac{1}{2m^2c^2}\,\frac{1}{r}\,\frac{d\Phi}{dr}\,(\underline{s}\,\underline{l})\quad. \tag{2.21}$$

For a Coulomb potential $\Phi(r) = -\dfrac{Ze^2}{r}$ we have

$$U_{so} \sim \frac{Z}{r^3}(\underline{s}\,\underline{l})\quad. \tag{2.22}$$

Because of $U_{so} \sim 1/r^3$ the spin-orbit interaction is strongest in the vicinity of the nucleus. This is possibly a reason why the muffin-tin approximation (see Sect. 2.4) works quite well in spin-polarized LEED. Because of $U_{so} \sim Z$ the spin-orbit interaction is largest for the heavy elements like W, Au, Hg, Pt, and rather weak for the light elements. This might possibly be a reason for Davisson and Germer's failure with Ni.

There is a classical argument for the spin-orbit interaction, which gives a plausible explanation for this effect. Consider an electron moving with the velocity $\underline{v}$ in an electric field $\underline{E}$ that has been generated by a centrosymmetric charge distribution. An observer in the rest frame of the electron sees, according to classical electrodynamics, a magnetic field

$$\underline{H} = -\frac{1}{c}(\underline{v} \times \underline{E}) + O(v^2/c^2)\quad.$$

The magnetic moment of the electron $\underline{\mu} = \dfrac{e}{mc}\,\underline{s}$ interacts with this magnetic field with the interaction energy

$$U' = -\underline{\mu}\,\underline{H}$$

with $\underline{p} = m\,\underline{v}$ we see that $U' \approx -(e/m^2c^2)\underline{s}(\underline{E} \times \underline{p})$. Identifying $\underline{s} = \underline{\sigma}\,\hbar/2$ shows that U' corresponds to the spin-orbit term in (2.20), except for a factor of two. This factor is due to our neglect of the Thomas precession /Kessler, 1976/.

The spin-orbit interaction thus corresponds to an additional scattering potential, the sign of which depends on whether an electron with spin-up or spin-down passes the scattering center on the right or on the left. This is illustrated in Fig. 2.1 for the Coulomb potential. The curve 'no spin' corresponds to the bare Coulomb potential. As the scattering potential becomes spin-dependent for trans-

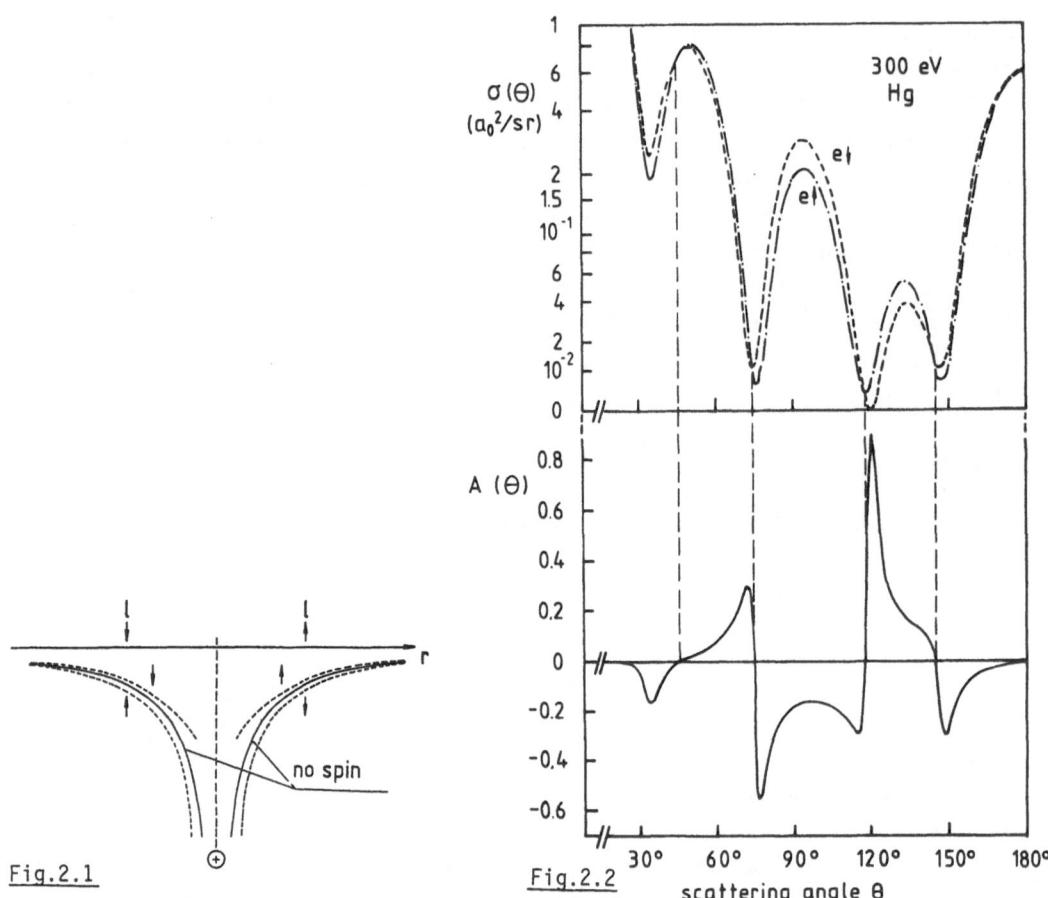

Fig.2.1

Fig.2.2

Fig. 2.1. Modifications to the Coulomb potential, due to spin-orbit coupling, as seen by a scattered electron with spin normal to the scattering plane. The curve 'no spin' indicates the pure Coulomb potential

Fig. 2.2. Differential cross section $\sigma(\theta)$ (upper panel) and asymmetry $A(\theta)$ (lower panel) for completely polarized electrons scattered elastically from free Hg atoms at 300 eV

versal polarized electrons, we should also expect the differential cross section to be different for spin-up and spin-down electrons. This is illustrated in Fig. 2.2 with an experimental result for the elastic scattering from Hg atoms. The differential cross section $\sigma(\theta)$ as a function of the scattering angle θ shows typical interference structure, determined by λ/R where R is the effective range of the potential and λ is the electron wavelength. For a given scattering angle the effective range depends on the sign and magnitude of the spin-orbit interaction. The latter is mostly small, which makes the shape of the differential cross section rather similar for both spin orientations. However, the scattering cross sections are slightly shifted with respect to each other on the θ axis. Therefore, the asymmetry A_{so}, defined by

$$A_{so}(\Theta) = \frac{\sigma^{\uparrow}(\Theta) - \sigma^{\downarrow}(\Theta)}{\sigma^{\uparrow}(\Theta) + \sigma^{\downarrow}(\Theta)} \qquad (2.23)$$

i.e. the normalized difference of spin-up and spin-down intensities at a given scattering angle Θ, shows pronounced structure mainly near absolute minima in the scattering cross section (see Fig. 2.2). The symmetry found in Fig. 2.1 immediately determines the asymmetry A_{so} when going to the complementary angle $-\Theta$. We see that the asymmetry $A_{so}(-\Theta)$ is obtained from $A_{so}(\Theta)$ by inversion with respect to $\Theta = 0°$:

$$A_{so}(-\Theta) = - A_{so}(\Theta) \quad . \qquad (2.24)$$

For example, the strong positive maximum at $\Theta \approx +125°$ is transformed into a strong negative maximum at $\Theta = -125°$. The important consequence is, that by means of spin-orbit coupling the measurement of the polarization of an electron beam can be reduced to two measurements of the intensity for complementary angles.

Relation Between Asymmetry and Scattering Amplitudes

The asymmetry $A_{so}(\Theta)$ is defined as the normalized difference of scattering cross sections. Therefore it can be expressed by means of scattering amplitudes for elastic scattering (see Kessler /1976/ for more details). By analogy with the non-relativistic scattering theory, the following ansatz is made for the asymptotic form ($r \to \infty$) of the scattered wave function:

$$\Psi' = \{\Psi'_n\} = a_n \, e^{ikz} + \frac{e^{ikr}}{r} \, u_n(\Theta,\phi) \qquad n = 1 \ldots 4 \quad .$$

The four components Ψ'_n are composed of plane waves travelling along z and spherical waves starting from the scattering center. The latter ones are modulated by functions $u_n(\Theta,\phi)$, which depend on the scattering angle, the electron energy, and the scattering potential. The differential scattering cross section then is given as in the non-relativistic case by

$$\sigma(\Theta,\phi) = \left(\sum_{n=1}^{4} |u_n(\Theta,\phi)|^2 \right) / \left(\sum_{n=1}^{4} |a_n|^2 \right) \quad . \qquad (2.25)$$

It has been shown /Kessler, 1976/ that for the case of elastic scattering only the two large components of the four-spinor need to be calculated. The incident electrons can be represented by a two-component spinor

$$\underline{\Psi} = \binom{a_1}{a_2} e^{ikz} = a_1 \binom{1}{0} e^{ikz} + a_2 \binom{0}{1} e^{ikz} \qquad (2.26)$$

where (2.3) has been used.

For incident electrons with spin parallel to momentum, after scattering the wavefunction is of the form

$$\Psi_{\pm\uparrow\uparrow} = a_1 \begin{pmatrix} e^{ikz} + f_1 \dfrac{e^{ikr}}{r} \\ 0 + g_1 \dfrac{e^{ikr}}{r} \end{pmatrix} \quad . \tag{2.27a}$$

For spin antiparallel to momentum:

$$\Psi_{\pm\uparrow\downarrow} = a_2 \begin{pmatrix} 0 + g_2 \dfrac{e^{ikr}}{r} \\ e^{ikz} + f_2 \dfrac{e^{ikr}}{r} \end{pmatrix} \quad . \tag{2.27b}$$

The total wavefunction after scattering is obtained by coherent superposition:

$$\underline{\Psi}' = \underline{\Psi}_{\pm\uparrow\uparrow} + \underline{\Psi}_{\pm\uparrow\downarrow} = \begin{pmatrix} a_1 \\ a_2 \end{pmatrix} e^{ikz} + \begin{pmatrix} a_1 f_1 + a_2 g_2 \\ a_1 g_1 + a_2 f_2 \end{pmatrix} \dfrac{e^{ikr}}{r} \quad . \tag{2.28}$$

The scattering amplitudes f_i and g_i then have to be determined by partial wave analysis. For a spherical potential Darwin /1928/ has given a general relation between the scattering amplitudes:

$$f_2(\Theta) = f_1(\Theta) \equiv f(\Theta)$$

$$g_2(\Theta,\phi) = -g_1(\Theta)e^{-2i\phi} \equiv -g(\Theta)e^{-i\phi} \quad . \tag{2.29}$$

The scattering amplitudes, expressed in the scattering phase shifts δ_ℓ^+, δ_ℓ^-, are

$$f(\Theta) = \frac{1}{2ik} \sum_{\ell=0}^{\infty} [(\ell + 1)(e^{2i\delta_\ell^+} - 1) + \ell(e^{2i\delta_\ell^-} - 1)] P_\ell(\cos\Theta) \tag{2.30}$$

$$g(\Theta) = \frac{1}{2ik} \sum_{\ell=1}^{\infty} [-e^{2i\delta_\ell^+} + e^{2i\delta_\ell^-}] P_\ell^1(\cos\Theta)$$

with $k = |\underline{k}| = \sqrt{2m_e E}/\hbar$ and $P_\ell(\cos\Theta)$ the Legendre polynomials.

By introducing the scattering amplitude $g(\Theta)$ the possibility of a spin flip during scattering is taken into account. Therefore $g(\Theta)$ is called 'spin-flip-amplitude'.

The quantities of physical importance, kinetic energy and scattering potential are implicitly present in the form of the scattering phase shifts $\delta_\ell^\pm$, which have to be determined in a theoretical numerical analysis.

Note that for the present case of eleastic scattering the phase shifts are real as the number of particles is conserved. Absorption effects can be taken into account by introducing complex phase shifts.

With the 'direct amplitude' $f(\Theta)$ and the spin-flip amplitude $g(\Theta)$ being known, the differential scattering cross section may be expressed in the scattering amplitudes by means of (2.25)

$$\sigma(\Theta,\phi) = \frac{|a_1 f - a_2 g e^{-i\phi}|^2 + |a_2 f + a_1 g e^{i\phi}|^2}{|a_1|^2 + |a_2|^2} \tag{2.31a}$$

12

or:

$$\sigma(\Theta,\phi) = (|f|^2 + |g|^2) \left[1 + A_{so}(\Theta) \frac{-a_1^* a_2 e^{i\phi} + a_1 a_2^* e^{-i\phi}}{i(|a_1|^2 + |a_2|^2)} \right] . \qquad (2.31b)$$

The asymmetry $A_{so}(\Theta)$ expressed in the scattering amplitudes is

$$A_{so}(\Theta) = i \frac{fg^* - f^*g}{|f|^2 + |g|^2} \quad \text{or} \quad A_{so}(\Theta) = -2 \frac{Im(fg^*)}{|f|^2 + |g|^2} . \qquad (2.32)$$

The asymmetry is seen to be a consequence of the interference between the direct amplitude f and the spin-flip amplitude g. In the special case of a beam complete-ly transversal polarized along the x axis ($a_1 = a_2 = 1$) we obtain

$$\sigma(\Theta,\phi) = (|f|^2 + |g|^2)[1 - A_{so}(\Theta)\sin(\phi)] . \qquad (2.33)$$

At fixed scattering angle Θ the differential scattering cross section is a func-tion of $\sin\phi$ with the extrema $\sigma_{max} \sim (1+A)$ and $\sigma_{min} \sim (1-A)$. The periodicity of this function of ϕ is 2π, which explains why Davisson and Germer looked for the wrong periodicity (they expected π, in analogy to light). In the case of purely longitudinal polarization ($a_1 = 1$, $a_2 = 0$ or $a_1 = 0$, $a_2 = 1$) the ϕ-dependent term in (2.31b) vanishes and there is no intensity variation as a function of the azi-muth ϕ. However, the *magnitude* of the cross section is still different from that of a 'no spin' electron because of the contribution from $|g|^2$. This fact is, for example, of importance for the calculation of electron backscattering in electron microscopy. A non-relativistic treatment usually yields too small backscatter co-efficients /Ichimura et al., 1980/.

We now show that the spin-orbit asymmetry A_{so} can be viewed as a vector quanti-ty. The spin-scattering matrix may be obtained from comparing (2.26) with (2.28) and using (2.29):

$$\underline{\underline{S}} = \begin{pmatrix} -f & -ge^{i\phi} \\ ge^{i\phi} & f \end{pmatrix} . \qquad (2.34)$$

We choose a coordinate system such that the primary electrons with wavevector $\underline{k}$ run along the +z axis. We define the normal $\underline{n}$ to the scattering plane by:

$$\underline{n} := \frac{\underline{k} \times \underline{k}'}{|\underline{k}| \times |\underline{k}'|} \qquad (2.35)$$

where $\underline{k}'$ is the wavevector of the scattered electron.

By inserting the spin-scattering matrix (2.34) into (2.18) we obtain for the polarization after scattering

$$\underline{p}' = \frac{2i(fg^* - f^*g) \underline{n}}{2(ff^* + gg^*)} .$$

Comparison with (2.32) shows:

13

$$\underline{P}'(\theta) = A_{so}(\theta) \; \underline{n} \; . \tag{2.36}$$

Thus we may redefine $\underline{A}_{so}: = A_{so} \; \underline{n}$ and we have $\underline{P}' = \underline{A}_{so}$.

This result shows two remarkable points:

- An unpolarized electron beam is polarized by scattering due to spin-orbit coupling. The polarization vector is normal to the scattering plane.
- The polarization $\underline{P}'$ after scattering an unpolarized beam is given by the same intensity asymmetry $\underline{A}_{so}$, that is obtained when scattering a polarized electron beam with $\underline{P}'$ normal to the scattering plane.

These results, in particular $\underline{P}' = \underline{A}_{so}$, have been deduced for electron-atom scattering only. Under certain conditions, however, they are also valid in diffraction from surfaces (see Chap. 4). For example, this is true if the scattering plane coincides with a mirror plane of the crystal. The equality of polarization and asymmetry is the basis for the self-calibration capability of a double scattering experiment. This aspect is treated in more detail in Chap. 3.

2.3 Exchange Interaction

The exchange interaction is a genuine quantum mechanical effect, without classical analogue. The Pauli exclusion principle for spin 1/2 particles requires, that the wavefunction describing a system of two or more particles must be antisymmetric with respect to the exchange of two particles. The spin-dependent forces arising as a consequence of this postulate are essentially of Coulombic nature. For the low energies considered here, they are much larger than e.g. the dipole-dipole interaction between two electrons. Therefore, in the non-relativistic limit, it is sufficient to use the Schrödinger equation instead of the Dirac equation, which was necessary to describe the spin-orbit interaction.

To illustrate some particular features of the exchange interaction let us consider the elastic scattering of two free electrons; for e.g. protons the arguments would be identical. Excluding spin-flip processes, the wavefunction of the two-particle system can be factorized into a spatial part and a spin function. The total spin of the system will be either 0 (singlet) or 1 (triplet). Within the center-of-mass frame the problem is equivalent to scattering from a central potential, and we may use the same ansatz as in the previous section for the scattered wavefunction, with one component only. All one has to do is to symmetrize properly according to the spin state of the system. If it is in a singlet state, the wavefunction will be antisymmetric with respect to the spin coordinates, hence symmetric with respect to the spatial coordinates. The symmetrized wavefunction for the singlet state then is

$$\Psi_s = V\{e^{ikz} + e^{-ikz} + [f_s(\theta) + f_s(\pi-\theta)] \frac{e^{ikr}}{r}\} \tag{2.37}$$

where V is a normalizing factor. The symmetrized scattering amplitude $f_s(\theta)$ is

$$f_s(\theta) = [f_s(\theta) + f_s(\pi-\theta)] \quad . \tag{2.38}$$

Hence, the differential scattering cross section for the singlet state is

$$\frac{d\sigma_s}{d\Omega} = |f_s(\theta) + f_s(\pi-\theta)|^2 \quad . \tag{2.39}$$

In the triplet state the spin function is symmetric and the spatial part has to be antisymmetric. The symmetrized scattering amplitude then is

$$f_t(\theta) = [f_t(\theta) - f_t(\pi-\theta)] \tag{2.40}$$

and the cross section for the triplet state is

$$\frac{d\sigma_t}{d\Omega} = |f_t(\theta) - f_t(\pi-\theta)|^2 \quad . $$

As explicitly spin-dependent forces are neglected, the scattering amplitudes of the triplet and the singlet states are equal:

$$f_s(\theta) = f_t(\theta) = f(\theta) \quad . \tag{2.41}$$

With the scattering amplitude for Coulomb scattering /Dawydow, 1967/ the final result in the center-of-mass system is:

$$d\sigma_s(\theta) = (\frac{e^2}{mv^2})^2 \left\{ \frac{1}{\sin^4(\frac{\theta}{2})} + \frac{1}{\cos^4(\frac{\theta}{2})} + \frac{2\cos[\lambda \ln(tg^2(\frac{\theta}{2}))]}{\sin^2(\frac{\theta}{2})\cdot\cos^2(\frac{\theta}{2})} \right\} d\Omega \tag{2.42a}$$

$$d\sigma_t(\theta) = (\frac{e^2}{mv^2})^2 \left\{ \frac{1}{\sin^4(\frac{\theta}{2})} + \frac{1}{\cos^4(\frac{\theta}{2})} - \frac{2\cos[\lambda \ln(tg^2(\frac{\theta}{2}))]}{\sin^2(\frac{\theta}{2})\cdot\cos^2(\frac{\theta}{2})} \right\} d\Omega \quad . \tag{2.42b}$$

with $\lambda = e^2/hv$ and v the relative velocity of the two electrons. It is seen that the two expressions differ only by the sign of the third term. This term is absent in a classical treatment and is due to the correlation of the particles by the symmetry requirements on the two-particle wavefunction upon exchange of the particles. The amplitude $f(\pi-\theta)$ is therefore called "exchange amplitude":

$$g_{ex}(\theta): = f(\pi-\theta) \quad . \tag{2.43}$$

The exchange amplitude g_{ex} must not be confused with the spin-flip amplitude g_{so} introduced above. In pure exchange interaction the spin orientations are conserved!

In Fig. 2.3 the result is given for the case of one electron being at rest in the laboratory frame while the other one has the energy $E_0 = 100$ eV. The dashed line indicates the shape of the cross section for scattering in the singlet state as a function of the laboratory scattering angle. The full line gives the ratio of the triplet cross section to the singlet cross section. For small and large scattering angles this ratio is 1, while for $\theta = \pm 45°$ the triplet cross section van-

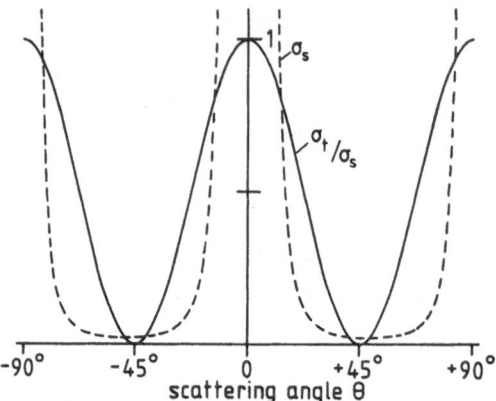

-90°   -45°   0   +45°   +90°
scattering angle θ

Fig. 2.3. Calculated singlet cross section σ_s (dashed line) for exchange scattering of free electrons as a function of scattering angle θ (transformed into the laboratory frame). The ratio σ_t/σ_s of triplet to singlet cross sections goes to zero near θ = ± 45°, giving rise to a large exchange asymmetry

ishes. At these angles the exchange asymmetry, defined as the normalized difference of the singlet and triplet cross sections approaches 1:

$$A_{ex}(45°) = \frac{\sigma_s - \sigma_t}{\sigma_s + \sigma_t} = 1 \qquad \text{since } \sigma_t(45°) = 0 \; . \tag{2.44}$$

In remarkable contrast to the spin-orbit asymmetry A_{so} the sign of the exchange
asymmetry does not change when going to the complementary angle:

$$A_{ex}(-\theta) = A_{ex}(\theta) \; . \tag{2.45}$$

This means that in exchange scattering there is no left-right asymmetry for electrons of opposite spin. Rather, there is an 'up-down' asymmetry at a given angle.
   We found the asymmetry in spin-orbit interactions to be caused by a spin-dependent additional interaction potential, see (2.21) or (2.22). In a similar way we
may introduce a spin-dependent 'exchange potential', which, depending on the relative orientation of the spins involved, either increases or decreases the effective scattering potential. This approach is discussed in more detail in Sect. 2.4.
   Throughout this discussion we did not consider the orientation of the electron
spins relative to their momentum or to the scattering plane. It is another characteristic feature of exchange scattering, that this is indeed of no importance, as
long as the non-relativistic limit is valid. All that is required is that the
spins be aligned relative to each other, be the polarization longitudinal or
transversal. This effect is due to the Coulombic nature of the exchange interaction, with the spin orientation only entering the wavefunction symmetry.
   The intensity asymmetry in exchange scattering of free electrons has been used
as a spin polarization detector (Møller scattering). If one scatters polarized

electrons from a magnetized iron foil, the scattered intensity in the vicinity of $\Theta_L = \pm 45°$ will show sizeable differences if the foil magnetization or the incident polarization is reversed. As Møller scattering is applicable for longitudinally polarized beams also, it has been much used in beta-decay studies. For more details see Kessler /1976/.

We have introduced the spin-orbit interaction and the exchange interaction in the simplest possible ways: scattering of an electron from a central potential or from another electron. In reality, already in electron-atom and even more so in electron-crystal scattering /Ravano et al., 1982/ things are more complex. In general, both kinds of interaction will be present and will interfere with each other. For example, consider an unpolarized electron beam scattered from aligned free atoms (e.g. selected by passing through an inhomogeneous magnetic field). Classically speaking, the scattering process may be decomposed into two steps: In the first step, the electrons may become polarized via spin-orbit coupling in scattering from the nucleus. The polarized electrons in the second step may undergo exchange scattering with an aligned electron of the outer shell. The final result will clearly be different from pure exchange scattering with an unpolarized primary beam. This was pointed out for one-electron atoms by Burke and Mitchell /1974/, and calculated for free Cs atoms by Walker /1974/. In a recent model calculation for multi-electron atoms ('magnetic muffin-tin atoms') (Ackermann and Feder /1984/, Tamura et al. /1984/) the interference effects were shown to be well discernible. For example, noticeable left-right asymmetry was obtained in addition to exchange asymmetry. The simple picture of a two-step scattering process is certainly oversimplified, but is perhaps not too far fetched if we take the different length scales into account. The radius of the spin-orbit coupling regime is roughly an order of magnitude smaller than the radius of the outer shell electron distribution giving rise to exchange interaction.

2.4 Diffraction of Spin-Polarized Electrons

In this section we discuss spin-dependent electron scattering from crystals, i.e. from a regular array of atoms. We first treat the single-scattering (or kinematic) approximation and we shall see that in this limit the spin polarization is very similar to that from the single atom. Beyond the kinematic approximation, i.e. with full multiple scattering, some general relations may be obtained from considering the structure of the spin-scattering matrix. Finally a brief description of the calculational procedures used in theory is given.

2.4.1 Kinematic Approximation

To simplest approximation the diffraction of low energy electrons (without spin) from a single crystal surface is equivalent to scattering of light from an optical grating. According to the principle of Huygens-Fresnel a two-dimensional periodic array of scatterers with basisvectors $\underline{a}$, $\underline{b}$ is excited by an incident plane wave with wavevector $\underline{k}$. They emit spherical waves (modulated by the scattering amplitudes u_λ), which shall not be scattered by neighbouring scattering centers. Under this condition, the "einhüllende" has the form of plane waves running to infinity. The condition for constructive interference is $\underline{k}'_\parallel = \underline{k}_\parallel + \underline{g}_\parallel$, where $\underline{k}'_\parallel$ ($\underline{k}_\parallel$) is the parallel component of the wavevector of the outgoing (ingoing) wave and $\underline{g}_\parallel$ an arbitrary reciprocal vector of the two-dimensional lattice. For a semi-infinite crystal a further periodicity normal to the surface is present, which requires a third Laue-condition to be valid for constructive interference. The diffraction condition then is $\underline{k} - \underline{k}' = \underline{g}$, where $\underline{g}$ is a reciprocal lattice vector of the 3-dimensional lattice ($\underline{a}$, $\underline{b}$, $\underline{c}$). There is thus a certain analogy to X-ray diffraction, with a major difference concerning the absorption of electrons. Due to the strong interaction of a kinetic electron with the electronic system of the solid (excitation of surface- and bulk-plasmons, electron-hole pair generation) there is a strong loss from the elastic channel. This process is approximately described by a 'mean free path' $\lambda_e(E)$ in an attenuation law of the form $N(z) \sim e^{-z/\lambda_e}$. $N(z)$ gives the number of electrons with energy E that did not loose energy after having travelled the path length z. Precisely speaking, the 'quasi-elastic' part is considered, i.e. those electrons that lost only small amounts of energy, e.g. by excitation of phonons or magnons, are treated as 'elastic'. This convention corresponds to the energy resolution of most LEED spectrometers. The mean free path λ_e is very similar for most materials and has a minimum around E = 50 eV, measured with respect to the Fermi-energy. For typical LEED energies (30 to 300 eV) λ_e is of the order of 0.5 nm /Seah and Dench, 1979/. This means that only a limited number of layers contributes to the diffracted wavefield. The Laue condition for the z axis is therefore relaxed, intensity maxima due to the z periodicity are often broad or of small intensity.

What should be expected for the spin polarization within the framework of kinematic theory? According to the assumption of negligible multiple scattering the amplitude at the detector is given by a summation of the direct and the spin-flip or exchange amplitudes from each atom separately. The periodicity of the array of scatterers enters via phase shifts. The scattered amplitudes at the detector are

$$F(\Theta) = \sum_j f_j \exp(i\underline{\Delta k}\ \underline{r}_j)$$
$$\Theta = \sphericalangle(\underline{k}, \underline{k}') \qquad\qquad (2.46)$$
$$G(\Theta) = \sum_j g_j \exp(i\underline{\Delta k}\ \underline{r}_j)$$

with $\underline{\Delta k} = \underline{k} - \underline{k}'$ and $\underline{r}_j$ the radius vector of the jth atom. The sum has to be taken over the area from where the scattering is coherent. For an unpolarized primary beam the intensity is

$$I(\Theta) = FF^* + GG^* = \sum_{j,l} (f_j f_l^* + g_j g_l^*)\exp[i\underline{\Delta k}(\underline{r}_j - \underline{r}_l)] \ . \qquad\qquad (2.47)$$

The polarization of the diffracted beam is according to (2.32)

$$P(\Theta) = i \frac{FG^* - F^*G}{|F|^2 + |G|^2} = i \frac{\sum_{j,l} (f_j g_l^* - f_j^* g_l) \exp[i\underline{\Delta k}(\underline{r}_j - \underline{r}_l)]}{\sum_{j,l} (f_j f_l^* + g_j g_l^*) \exp[i\underline{\Delta k}(\underline{r}_j - \underline{r}_l)]} \quad . \tag{2.48}$$

For the case of an elemental crystal we have $f_l = f_j = f$ and $g_j = g_l = g$ and we find from the double sum:

$$I(\Theta) = (|f|^2 + |g|^2) \sum_{l,j} \exp[i\underline{\Delta k}(\underline{r}_j - \underline{r}_l)] \tag{2.49a}$$

$$P(\Theta) = i \frac{(fg^* - f^*g) \sum \cdots}{(|f|^2 + |g|^2) \sum \cdots} = i \frac{fg^* - f^*g}{|f|^2 + |g|^2} \quad . \tag{2.49b}$$

The lattice sum $\sum_{l,j} \exp[i\underline{\Delta k}(\underline{r}_j - \underline{r}_l)]$ which determines maxima and minima of the intensity is the same in nominator and denominator of the expression (2.49b) for the polarization. Thus we find the remarkable result that within the kinematic approximation the spin polarization for the crystal is the same as for the single atom. In spin-orbit interaction the polarization vector $\underline{P}$ is normal to the scattering plane, as well as the asymmetry vector $\underline{A}$, and $\underline{P} = \underline{A}$ remains valid.

Likewise, in exchange scattering the measured asymmetry or polarization is determined only by the magnetic atomic scattering potential.

Up to now we assumed the scatterers to form a rigid lattice. In reality, even at T = 0 K, the atoms oscillate about their equilibrium positions, which is in kinematic theory taken into account by the so-called Debye-Waller factor. It is assumed that the atomic displacements are isotropic and uncorrelated. For the phonon density of states a Debye model is assumed. This is not a very sophisticated model of the thermal vibrations, but has been used with good success in LEED /Pendry, 1974/. Since the atomic displacements are slow compared to typical scattering times, the atoms are assumed to be at rest in the scattering process (Born-Oppenheimer approximation). The displacements then lead to a reduction of the scattering amplitudes that interfere constructively. When averaging over the time for a measurement the scattering amplitudes are reduced by the factor

$$\exp[-\frac{1}{2} \langle (\underline{\Delta k}\underline{\Delta r}_j)^2 \rangle_t] \quad .$$

$\underline{\Delta k}$ represents the momentum transfer and $\underline{\Delta r}_j$ indicates the displacement from the average position $\underline{r}_j$. The scattered intensity I for one atom per unit cell at temperature T then is proportional to the 'Debye-Waller factor':

$$I \sim \exp(-\frac{3|\underline{\Delta k}|^2 T}{mk_B \Theta_D^2})$$

where k_B is Boltzmann's constant and Θ_D is the Debye temperature. At a surface the mean square displacement may be larger than in the bulk, so that the surface Debye temperature frequently is lower than the bulk Debye temperature. When determining the polarization from (2.49b) we note that the Debye-Waller factors in the nominator and the denominator cancel each other. Consequently, in kinematic theory the polarization is *independent of temperature*.

Let us briefly summarize the consequences of the kinematic approximation:

1) In contrast to the intensity, the polarization yields no information on the crystallografic structure of the surface.
2) Intensity and polarization are 'decoupled': The polarization is determined by the properties of the single atom while the intensity is determined by coherent summation of the scattered amplitudes. Thus, in principle a high polarization may be found simultaneously with high intensity. Such a combination would be very useful for a polarization detector. We will see in Chap. 3 how far this is true.
3) The polarization is independent of temperature. Therefore temperature dependent changes of the vibration amplitudes cannot be detected. Because of the insensitivity to structural properties other temperature dependent phenomena like lattice expansion or reconstruction also cannot be detected.
4) In scattering from magnetic surfaces the exchange asymmetry is a measure of the effective magnetic moment per atom averaged over the depth of information of the experiment.

From the above it might appear at first glance that the study of polarization phenomena is not particularly interesting, except perhaps for magnetic materials. The following discussion, however, will show that the single scattering limit is only rarely fulfilled and that multiple scattering is of major importance in almost all cases. The 'negative' statements are then converted into their opposite!

Under what conditions should we expect a single scattering approximation to be valid? Returning to the Fresnel-Huygens picture we would expect the contribution from neighbouring atoms to the total intensity to be small if (i) the scattering cross section is small or, (ii) if the absorption is strong. The scattering cross section is comparable to the geometric size of atoms, i.e. of the order of 10^{-16} cm^2 for medium to heavy atoms. On the other hand there are LEED results, e.g. for solid Xe crystals (Z = 54) /Ignatiev et al., 1971/, that are in excellent agreement with kinematic theory. The reason is, that it is neither cross section nor absorption alone that counts, but rather the ratio of the two effects. Multiple scattering becomes negligible if the amplitude of the scattered wave is so strongly damped on its way to the neighbouring atom that the excited amplitude is negligible. Thus we may have a kinematic situation when the mean free path is small or when the lattice parameter is large. The mean free path is similar for all materials at a given energy, being essentially determined by the electron density. However, small changes of the lattice constant may have large effects as the damping depends exponentially on the path length. The lattice constant of Xe single crys-

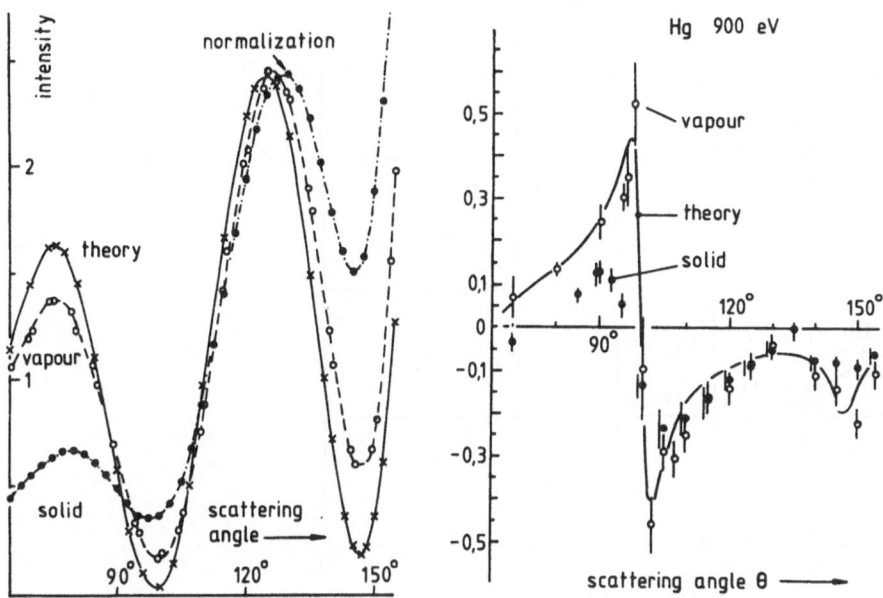

Fig. 2.4. Elastic scattering from solid and gaseous mercury at 900 eV (angular resolution Δθ).
Left hand side: angular distributions from a solid Hg target (— • —) with Δθ = 3°; from free Hg atoms (— o —) with Δθ = 1°; theory for free atoms (— x —).
Right hand side: spinpolarization from solid Hg (♦) with Δθ = 8°; from Hg atoms (φ) with Δθ = 6°; and theory (———) scaled to Δθ = 8°

tals is as large as 6.15 Å (von der Waals bonding!) which causes strong damping and hence a kinematic behaviour.

A different situation arises when the scatterer is strongly disordered. We would expect interference effects to be washed out, resulting in a broad structureless background onto which the atomic scattering cross section is superimposed. This is not kinematic scattering in the proper sense but rather resembles scattering from a liquid (Schilling and Webb /1970/, Unertl /1979/). An example is given in Fig. 2.4, showing elastic scattering cross sections and polarization from gaseous and solid mercury /Eckstein, 1970/. We note that the minima in the differential cross section are filled up for solid Hg while the general structure is preserved. The polarization structure around θ = 100° is damped but well visible, which points towards the contribution of an unpolarized background. Similar observations have also been made at 100-500 eV for the differential cross section. A characteristic feature in scattering from liquid-like or amorphous surfaces is, that intensity and polarization depend on the scattering angle only, irrespective of the polar and azimuthal angles of incidence of the primary beam, if a correction for the absorption is applied. This gives an experimental hint, whether a kinematic point of view may reasonably be adopted in the analysis of data. For example, these features have been observed in exchange scattering from ferromagnetic metallic glasses /Pierce et al., 1982/ which have a frozen-liquid structure.

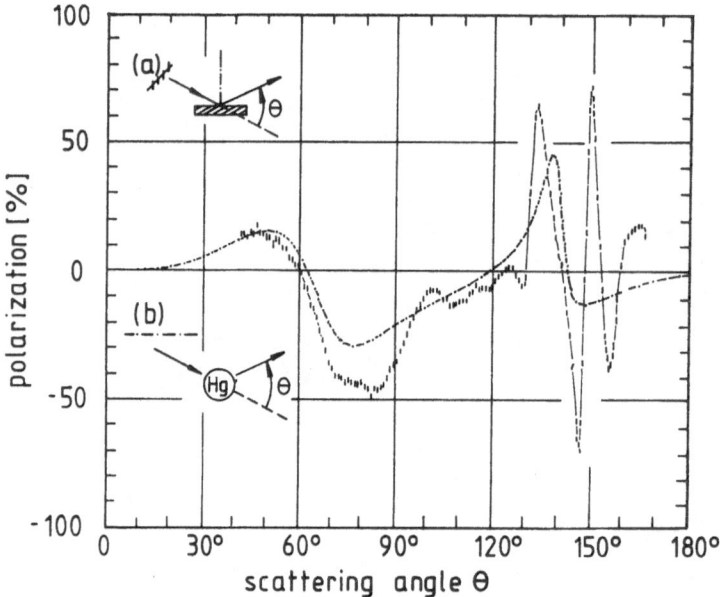

Fig. 2.5. Comparison of spin-polarization effects in electron-atom scattering and in low energy electron diffraction. Curve (a) (ııııı) shows the result for the specular beam from Au(110) as a function of the scattering angle Θ. Curve (b) (-•-•-) is the calculated result for Hg atoms

Though the presence of the background of unknown magnitude and polarization poses problems, there is hope to determine the magnetic atomic scattering potentials by this technique.

Is the atomic behaviour totally obscured in diffraction from a crystal or is it just modified? An example where the atomic features are relatively well seen is shown in Fig. 2.5. In this figure the polarization of the (0,0) beam from Au(110) at 70 eV as a function of the scattering angle is compared to a theoretical curve for free Au atoms /Müller, 1979/. There are features reminiscent of the free atom polarization but also strong deviations around Θ = 150° which are due to multiple scattering.

While the previous examples show atomic-like or liquid-like behaviour, eventually modulated by multiple scattering effects, the example of Fig. 2.6 shows structures that are exclusively due to multiple scattering. In this experiment the energy and the wavevectors $\underline{k}$ and $\underline{k}'$ were kept constant ((0,0) beam, E = 100 eV) while the crystal was turned about the surface normal ('rotation diagram'). If the kinematic approximation were valid, intensity and polarization should show constant values, independent of the azimuthal angle φ, as the kinematic conditions remain constant. In the experiment and in dynamical theory both show pronounced structure with a fourfold symmetry which corresponds to that of the W(001) surface. The very existence of these structures is a proof for the importance of multiple scattering for intensity and polarization /Kirschner and Feder, 1979/.

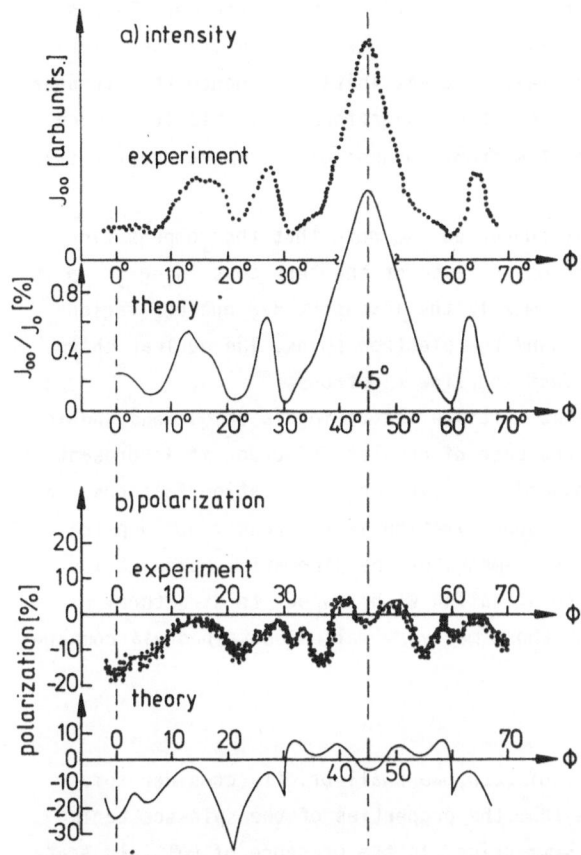

Fig. 2.6. Rotation diagrams for the specular beam from W(001) (E = 100 eV, Θ = 47.5°, φ variable) for a) intensity and b) spinpolarization. The dotted curves are experimental, the full lines theoretical results

One might be tempted to correlate the observation of pronounced multiple scattering with the atomic number Z of those materials that show strong spin-orbit coupling. For example, it is well known from electron microscopy that the backscattering cross section increases strongly with the atomic number. The experience acquired in LEED, however, shows that the diffraction from light elements like Al is as much dominated by multiple scattering as that from heavy metals. The reason for this relative insensitivity with respect to Z at typical LEED energies is given by the pseudopotential compensation theorem /Pendry, 1974/. Briefly, it states that an attractive potential, strong enough to support bound states, behaves with respect to scattering of slow particles as if all its energy had been consumed by this effort. This behaviour may be made plausible by a classical argument. A slow electron spends a relatively long time in the outer zones of the atom, where the strong potential of the nucleus is screened by the core electrons. When approaching the nucleus, it is strongly accelerated and passes by the nucleus quickly. The larger the atomic number, the stronger the acceleration but the shorter the residence time. Therefore the stronger attraction is largely compensated. The relative change of velocity is the larger, the smaller it was initially, hence the stronger is the compensation effect. High energy electrons therefore feel relatively more of the nucleus than slow ones, so that the atomic number gains importance for fast electrons.

As far as spin-orbit interaction is concerned, we note that the compensation theorem is not valid for the polarization. Because of the term containing $\underline{E} \times \underline{p}$ in the spin-orbit operator of (2.20) it is mainly the less-screened nuclear region with high field strength that turns around the electron spins. The nuclear charge Z therefore is equally important for fast and slow electrons.

Concluding this section we state that multiple scattering is ubiquitous and in general cannot be neglected. Even in the case of complete disorder it is present, though it does not give rise to pronounced intensity or polarization features. In diffraction from crystals the kinematic approximation is rigorously not applicable. As a consequence, the 'negative' statements of the kinematic approximation concerning the insensitivity of spin polarization with respect to structural parameters is obsolete. We will see that those may very well be obtained via comparison of theory and experiment.

2.4.2 Symmetry Properties

Before turning to the dynamical theory of LEED, we shall briefly consider some general conclusions that may be drawn from the properties of the spin-scattering matrix for particular geometries and symmetries. In the presence of multiple scattering the symmetry of the total system 'crystal plus primary beam plus detector' has to be considered, including mirror-, inversion- and time-reversal symmetries. This has consequences for the relation between the polarization vector $\underline{P}$ and the asymmetry vector $\underline{A}$ and their components.

Time Reversal Symmetries

Elastic scattering of 'spin-less' particles is time-reversal invariant. This is shown as follows: If $\psi(t)$ is the solution of the non-relativistic Schrödinger equation, the solution $\psi(-t)$ at time $-t$ is obtained by applying the operator T_0, which transforms ψ into its complex conjugate /Messiah, 1969/:

$$\psi(-t) = T_0\psi(t) = \psi^*(t) \ . \tag{2.50}$$

The intensity measured experimentally is given by the square of the wave function. It is thus the same for the time reversed state. If a scattering experiment is done with 'spin-less' primary particles, the interchange of source and detector does not change the measured intensity. For the specular reflected beam in 'ordinary' LEED this has an interesting consequence (Woodruff and Holland /1970/, Lagally et al. /1971/): The interchange of source and detector is equivalent to a rotation of the crystal by 180° about the surface normal. Therefore, a rotation diagram for a surface with 1-fold symmetry will show a 2-fold symmetric intensity structure. A 2-fold symmetric surface also shows 2-fold symmetric intensity. Thus a distinction from a 1-fold symmetric surface is not possible. Correspondingly a 3-fold symmetric surface yields 6-fold symmetric intensity, and a 4-fold symmetric surface yields a 4-fold symmetric intensity. When the electron spin is explicitly taken into account, these degeneracies are partly removed.

The time reversal operator for spin 1/2 particles is /Messiah, 1969/

$$T = i\sigma_y T_o \tag{2.51}$$

where σ_y is the y component of the spin operator $\underline{\sigma}$. This operator transforms the spin function χ into its time-reversed state χ'

$$\chi \xrightarrow{T} \chi' = -i\sigma_y \chi^* \quad .$$

Example: $\chi = \binom{1}{0} \xrightarrow{T} \chi' = -i\begin{pmatrix} 0 & -i \\ i & 0 \end{pmatrix}\binom{1}{0}^* = \binom{0}{1} \quad .$

It is also easy to show for the spin operator $\underline{\sigma}$ that

$$\underline{\sigma} \xrightarrow{T} -\underline{\sigma} \tag{2.52}$$

when using the commutation rules (2.1) and the self-adjunctness of the spin matrices. One therefore may say "time reversal reverses the spin". For this reason a time reversal operation with a magnetic crystal means interchanging source and detector *and* reversing all magnetic fields. With a non-magnetic crystal and polarized electrons the degeneracy of 1- and 2-fold and of 3- and 6-fold symmetry is removed. A surface of 4-fold symmetry like W(001) however, must show 4-fold symmetry of intensity and polarization. This is in agreement with Fig. 2.6, where mirror symmetry lines exist at $\phi = 0°$ and $\phi = 45°$. (Slight distortions are due to a small misalignment of rotation axis and surface normal.)

Spatial Symmetries

In contrast to scattering from atoms, in diffraction from surfaces polarization components lying in the scattering plane may appear due to multiple scattering, in particular longitudinal polarization, even when spin-orbit interaction is the polarization mechanism. As is well known the appearance of longitudinal polarization in β decay is a proof for parity violation in this process. In our case, the appearance of longitudinal polarization is not due to a parity violation. Rather, longitudinal components are allowed because the Hamilton operator of the system 'half-space with surface' is by itself not invariant with respect to the parity operation $\underline{r} \rightarrow -\underline{r}$.

However, if the scattering plane coincides with a mirror plane of the crystal, there is no more a component of the $\underline{P}$ vector parallel to the scattering plane.

This is shown as follows: The result of the scattering experiment must be identical to that of the mirror experiment. As the polarization $\underline{P}$ is an axial vector, it transforms with respect to the mirror operation like the vector product of two polar vectors. Therefore each component of the vector product parallel to the mirror plane changes sign upon mirror operation. Thus the experiment is only invariant with respect to the mirror operation if the parallel components of the axial vector vanish identically, i.e. if the polarization vector is normal to the mirror plane. This corresponds exactly to the scattering from atoms and there as well as here we find $\underline{P} = \underline{A}$.

Further general relations between $\underline{A}$ and $\underline{P}$ for the case of spin-orbit coupling were given by Feder /1980/, Dunlap /1980/ and Feder and Kirschner /1981a/. The spin state of a diffracted beam with wavevector $\underline{k}'$ may be represented by the two-component spinor $v_{\underline{k}'}$. This is obtained from the spinor $u_{\underline{k}}$ of the incident beam by multiplication with the spin-scattering matrix $\underline{\underline{S}}_{\underline{k}\underline{k}'}$. According to (2.18) for an unpolarized primary beam $\underline{P}_0 = 0$ the diffracted beam aquires the polarization $\underline{P}_{-\underline{k}\underline{k}'}$. On the other hand, we find the intensity asymmetry $A_{\underline{k}\underline{k}'}(\underline{P}_0)$ in the diffracted beam if the polarization $\underline{P}_0$ of the primary beam is reversed. Thus we may define a generalized asymmetry vector $\underline{A}_{-\underline{k}'\underline{k}}$ by means of

$$\underline{A}_{-\underline{k}\underline{k}'}\underline{P}_0 = A_{\underline{k}\underline{k}'}(\underline{P}_0) \quad . \tag{2.53}$$

This definition also gives the prescription for the determination of the asymmetry vector. For a given $(\underline{k},\underline{k}')$ the orientation of $\underline{P}_0$ in space is varied until the intensity asymmetry $A_{\underline{k}\underline{k}'}$ upon reversal of $\underline{P}_0$ is maximum. The orientation of $\underline{P}_0$ then gives that of $\underline{A}_{-\underline{k}\underline{k}'}$ (compare (2.36)). Note, however, that in general the two vector quantities $\underline{P}_{-\underline{k}\underline{k}'}$, the polarization after scattering of an unpolarized primary beam, and the generalized asymmetry $\underline{A}_{-\underline{k}\underline{k}'}$ obtained from scattering a totally polarized beam, are *not* equal:

$$\underline{A}_{-\underline{k}\underline{k}'} \neq \underline{P}_{-\underline{k}'\underline{k}'} \quad .$$

Even if we do not explicitly know the elements of the spin-scattering matrix $\underline{\underline{S}}_{\underline{k}\underline{k}'}$, symmetries in space and time may lead to particular relations between $\underline{A}_{-\underline{k}\underline{k}'}$ and $\underline{P}_{-\underline{k}\underline{k}'}$. The spin-scattering matrix, being a (2x2) matrix, can be expanded in the Pauli matrices which have the property of completeness:

$$\underline{\underline{S}}_{\underline{k}\underline{k}'} = a\underline{1} + \underline{b}\underline{\underline{\sigma}} = \begin{pmatrix} a + b_x\sigma_x \\ a + b_y\sigma_y \\ a + b_z\sigma_z \end{pmatrix} \qquad \text{with } a, b_\nu \text{ complex} \quad . \tag{2.54}$$

With this representation, using (2.17) for the polarization $\underline{P}_{\underline{k}\underline{k}'}$ after scattering an unpolarized beam $(\underline{P}_0 = 0)$, the polarization $\underline{P}_{\underline{k}\underline{k}'}$ can be expressed in the expansion coefficients $a, \underline{b}$:

$$\underline{P}_{-\underline{k}\underline{k}'} = \frac{a\underline{b}^* + a^*\underline{b} - i(\underline{b}^* \times \underline{b})}{|a|^2 + |\underline{b}|^2} \qquad \begin{array}{l} a = a(\underline{k},\underline{k}') \\ \underline{b} = \underline{b}(\underline{k},\underline{k}') \end{array} \quad . \tag{2.55a}$$

Analogously we find for $A_{-kk'}$ using (2.53)

$$A_{-kk'} = \frac{ab^* - a^*b + i(b^* \times b)}{|a|^2 + |b|^2} .$$ (2.55b)

The two vectors $A_{-kk'}$ and $P_{-kk'}$ obviously are in general not equal because of the sign change in the nominator. For the magnitude, however, the sign is of no importance and we find the general rule

$$|A_{-kk'}| = |P_{-kk'}| .$$ (2.56)

This means that the *length* of the two vectors for identical scattering conditions (k,k') is equal, though the orientation in space may be different.

Next let us consider time reversal. It corresponds to the interchange of source and detector (plus, eventually, reversal of the magnetic fields):

$$k \overset{T}{\to} -k' \qquad \text{and} \qquad k' \overset{T}{\to} -k .$$

If the Hamiltonian operator of the total system is invariant with respect to time reversal, so must be the spin-scattering matrix:

$$\underset{=}{S}_{kk'} = \underset{=}{S}_{-k'-k} .$$

As, on the other hand we know from (2.52) that $\sigma \to -\sigma$ upon time reversal, we obtain:

$$a(k,k') \overset{T}{\to} a(-k',-k) \qquad \text{and} \qquad b(k,k') \overset{T}{\to} -b(-k',-k) .$$

Therefore, in the nominator of the expressions (2.55a) and (2.55b) the sign reverses upon time reversal and it follows:

$$A_{kk'} \overset{T}{\to} -P_{-k'k'} \qquad \text{and} \qquad P_{kk'} \overset{T}{\to} -A_{-kk'} .$$ (2.57)

The asymmetry is thus equal to the negative of the polarization after time reversal, and the polarization is equal to the negative time-reversed asymmetry.

In the case of the specular beam from a non-magnetic crystal with a 2-fold symmetry axis C_2 without mirror plane there is time inversion symmetry and rotation symmetry. For the coefficients a and b one obtains:

$$a \to a \qquad b_x \to -b_x \qquad b_y \to -b_y \qquad b_z = 0 .$$

It follows, that the components of $A_{-kk'}$ and $P_{-kk'}$ parallel to the surface are equal, and that the normal components are antisymmetric:

$$A_\parallel = P_\parallel \qquad \text{and} \qquad A_\perp = -P_\perp .$$

The parallel components change sign upon time reversal, while the normal components do not. The magnitudes $|A|$ and $|P|$ are equal. The definition of the various components of P is visualized in Fig. 2.7. The above result is also valid for the

27

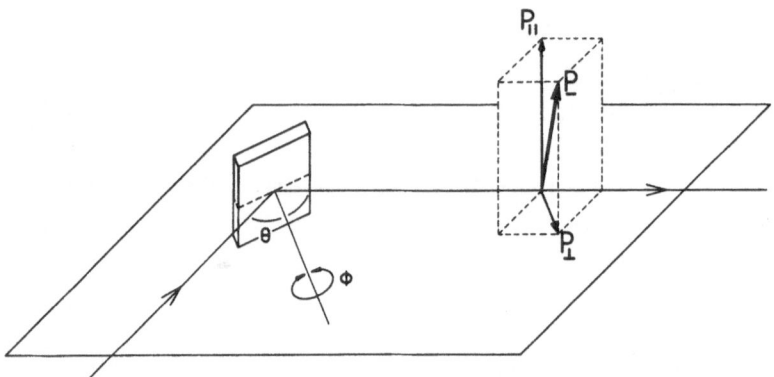

Fig. 2.7. Definition of the 'parallel component' $P_\parallel$ and the 'normal' component $P_\perp$ of the polarization vector $\underline{P}$

higher symmetries C_4 and C_6. In the case of a 3-fold rotation axis, rotation diagrams of the (0,0) beam show a 6-fold symmetry in intensity and the magnitudes of $\underline{A}$ and $\underline{P}$, $A_\perp$ and $P_\perp$, however, exhibit only a 3-fold symmetry.

There is a special case of some importance: normal incidence and normal reflection, i.e. the (0,0) beam at $\Theta = 0°$. In this case $\underline{k}' = -\underline{k}$, so that time inversion yields $\underline{k} \to -\underline{k} \equiv \underline{k}$. For the coefficients $\underline{b}$ in (2.54) follows

$$\underline{b}(\underline{k},\underline{k}') = \underline{b}(\underline{k},-\underline{k}') \overset{T}{\to} -\underline{b}(-\underline{k}',-\underline{k}) = -\underline{b}(\underline{k},-\underline{k}) \quad .$$

This is only fulfilled for $\underline{b} \equiv 0$, which means (using (2.55a,b))

$$\underline{P} = \underline{A} \equiv 0 \quad .$$

An unpolarized beam cannot be polarized from a non-magnetic crystal at normal incidence, and a polarized beam does not cause an intensity asymmetry. Rotation diagrams vanish identically for $\Theta \to 0$. Note, however, that beams of higher order, if present, may be polarized, also for normal incidence, as for those beams $\underline{k}' \neq -\underline{k}$.

For magnetic crystals, in the absence of spin-orbit coupling, only the component $\underline{b}$ parallel to the magnetization axis is non-zero. Consequently $\underline{P}$ and $\underline{A}$ are always equal and parallel to the magnetization axis. Time reversal (including reversal of the magnetic fields) changes the sign of $\underline{P}$ and $\underline{A}$. In the presence of spin-orbit coupling complicated interferences arise. There are, however, two par-

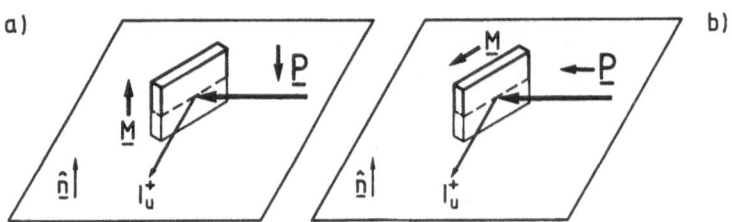

Fig. 2.8. 'Perpendicular' (a) and 'planar' (b) scattering geometries

28

ticular scattering geometries that allow for a good decoupling of exchange inter-action and spin-orbit interaction in the experiment. In the 'perpendicular geom-etry', see Fig. 2.8a, the polarization $\underline{P}$ of the primary beam and the magnetization $\underline{M}$ of the sample both are normal to the scattering plane which is chosen to coin-cide with a mirror plane of the crystal. For a diffracted beam there are four dif-ferent intensities $I_\sigma^\pm$ to be measured, depending on the orientation of $\underline{M}$ and $\underline{P}$ with respect to the normal to the scattering plane:

$$I_u^+(\underline{P} \uparrow\uparrow \underline{n};\ \underline{M} \downarrow\uparrow \underline{n}),\ I_d^+(\underline{P} \downarrow\uparrow \underline{n};\ \underline{M} \downarrow\uparrow \underline{n})\ ,$$

$$I_u^-(\underline{P} \uparrow\uparrow \underline{n};\ \underline{M} \uparrow\uparrow \underline{n}),\ I_d(\bar{P} \downarrow\uparrow \underline{n};\ \underline{M} \uparrow\uparrow \underline{n})\ . \tag{2.58}$$

From these intensities one may form two experimental asymmetries:

$$A^+:\ = \frac{1}{P_{-o}} \cdot \frac{I_u^+ - I_d^+}{I_u^+ - I_d^+} \qquad \text{for the magnetization antiparallel to } \underline{n}$$

$$\tag{2.59}$$

$$A^-:\ = \frac{1}{P_{-o}} \cdot \frac{I_u^- - I_d^-}{I_u^- + I_d^-} \qquad \text{for the magnetization parallel to } \underline{n}\ .$$

From these we obtain the approximate exchange- and spin-orbit asymmetries /Alvara-do et al., 1982/:

$$A_{ex} = \frac{1}{2}(A^+ - A^-) + \ldots$$

$$\tag{2.60}$$

$$A_{so} = \frac{1}{2}(A^+ + A^-) + \ldots\ .$$

The correction terms are of order $A_{ex} \cdot A_{so}$, as they arise from the interference of the exchange and spin-flip amplitudes. In the case of Ni they were found to be negligible /Alvarado et al., 1982a/, though this need not be the case in general. We mentioned already that, due to the mixed terms, even for an unpolarized primary beam an asymmetry A_u may be observed upon reversal of the magnetization:

$$A_u = \frac{I_u^+ + I_d^+ - I_u^- - I_d^-}{I_u^+ + U_d^+ + I_u^- + I_d^-} = \frac{I^+ - I^-}{I^+ + I^-}\ .$$

This can be made plausible by decomposing the scattering process into two steps. In the first step the unpolarized beam is polarized by spin-orbit interaction, with the polarization normal to the scattering plane. The polarized electrons then undergo exchange scattering with the target electrons as $\underline{M}$ is also normal to the scattering plane. This heuristic picture of course only serves for illustration. An appropriate relativistic diffraction theory has recently been developed by Ta-mura et al. /1984/. There it has also been pointed out that the interference ef-fects could be used for LEED studies of magnetic surfaces without polarized pri-mary electrons or a polarization detector. This approach might be viable for heavy ferromagnets such as Gd (see Sect. 5.1.1).

In the 'planar geometry' (see Fig. 2.8b), the spin-orbit induced polarization of an initially unpolarized beam would be orthogonal to the magnetization $\underline{M}$. Therefore, there is no spin-dependent interaction, the scattered intensity does not change upon magnetization reversal, and the asymmetry A_u vanishes. If the primary beam is polarized in the scattering plane, the effective polarization is the component of $\underline{P}$ parallel to $\underline{M}$. The existence of a spin-flip amplitude in spin-orbit interaction might cause this component not to be the same as the one in the free electron beam as the polarization vector might be rotated in the scattering plane. For light ferromagnets like Fe and Ni these effects are expected to be small, however, and the planar geometry may be used to isolate the exchange interaction in electron scattering or inverse photoemission experiments (cf. Sects 5.1 and 5.3).

2.4.3 Dynamical Theory

The basic situation in a LEED experiment is sketched in Fig. 2.9. An ideally mono-chromatic and parallel electron beam, represented by a plane wave with wavevector $\underline{k}$, impinges onto a semi-infinite ideal crystal. There is a certain number of elastically backdiffracted plane wave states with wavevector $\underline{k}'$ and a number of waves travelling into the crystal with wavevector $\underline{q}$. The theoretical task is to determine the intensity and spin polarization of the diffracted beams for a given energy, primary polarization, and magnetic and geometric structure of the crystal and its surface. This means that the wavefield of ingoing and outgoing vacuum states, matched across the surface to the outgoing crystal states has to be calculated. Let us consider the states involved in more detail. For the outgoing plane wave states the parallel component $\underline{k}_\parallel$ of the wavevector $\underline{k}$ is conserved modulo $\underline{g}_\parallel$, where $\underline{g}_\parallel$ is a reciprocal lattice vector of the surface lattice.

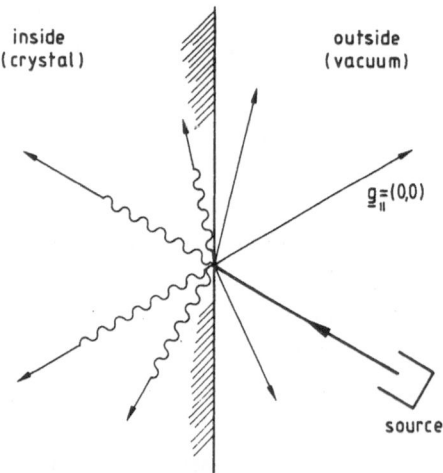

inside
(crystal)

outside
(vacuum)

$\underline{g}_\parallel = (0,0)$

source

Fig. 2.9. The "LEED-state". Ingoing and outgoing free electron spinors are matched at the interface with respect to phase and amplitude to outgoing Bloch spinors of the crystal

As we consider elastic scattering only, there must be energy conservation:

$$\frac{2mE}{\hbar^2} = k_{\perp}'^2 + |\underline{k}_{\parallel} + \underline{g}_{\parallel}|^2 \quad \text{for beams in the vacuum } (\underline{k}' = (\underline{k}_{\parallel}, k_{\perp}')) \qquad (2.61a)$$

$$\frac{2m(E-V_0)}{\hbar^2} = q_{\perp}^2 + |\underline{k}_{\parallel} + \underline{g}_{\parallel}|^2 \quad \text{for Bloch states in the crystal .} \qquad (2.61b)$$

The momentum transfer due to the reflection of electrons is absorbed by the crystal as a whole. The addition of the 'inner potential' V_0 takes into account that there is a potential step at the surface that prevents the electrons from leaving the crystal. The electrons are described by wavefunctions that have to be eigenfunctions at the appropriate Hamilton operator inside and outside of the crystal. In the crystal they are represented by Bloch states, which can be expanded into plane waves according to the Fourier representation of the periodic potential:

$$\psi_{\underline{q}}(\underline{r}) = e^{i\underline{q}\underline{r}} w_{\underline{q}}(\underline{r}) = e^{i\underline{q}\underline{r}} \sum_{\underline{g}} \alpha_{\underline{q}-\underline{g}} e^{-i\underline{g}\underline{r}} . \qquad (2.62)$$

The relation between energy E and wavevector $\underline{q}$ of Bloch states is given by the band structure $E(\underline{q})$. Among all Bloch states of the crystal only those are excited for which energy and parallel momentum $q_{\parallel}$ agree with those of the free electron wave, and for which the normal component $q_{\perp}$ satisfies (2.61b). Of these Bloch states only those are excited that represent electrons travelling into the crystal, i.e. those with a group velocity $\partial E/\partial q_{\perp}$ pointing away from the surface. Therefore, a high reflectivity is to be expected if there are only 'wrong' Bloch states or if there are no Bloch states at all, i.e. for a gap in the band structure. This case mainly appears at low energies around 10 eV, at higher energies gaps are rare.

   Allowed Bloch states do not have infinite extension, as the 'hot' electrons in the crystal have a finite mean free path. They are thus described by damped Bloch states, characterized by a complex wave vector. The excited states consequently have to be chosen from the complex band structure. Each $\underline{q}$ then has real and imaginary components, which means that there are allowed states in band gaps and that gaps in the real part of the band structure [i.e. $E(\text{Re}(\underline{q}))$] are closed /Pendry, 1974/.

   When expanding the Bloch states in plane waves, cf. (2.62), we have in principle to sum over all possible reciprocal lattice vectors $\underline{g}$, i.e. over an infinite manifold. Fortunately, for sufficiently large $\underline{g}$ the normal component $q_{\perp}$ becomes imaginary, even without damping. The larger the imaginary part, the faster the decay and the smaller the weight of the corresponding plane wave in the Fourier-analysis. In principle, therefore, convergence can always be reached, but it depends on the particular case, how many 'beams', i.e. reciprocal lattice vectors have to be taken into account. It has been found that for energies up to 150 eV about 40 reciprocal lattice vectors are sufficient. This means that the number of

diffracted beams is always much smaller than the total number of beams to be considered.

If we now had determined the complex band structure and with it the relevant
Bloch states, the LEED problem would still not have been solved. We have to match
the amplitudes and phases of all ingoing and outgoing waves across the surface.
The matching condition is, that the wavefunction and its derivative have to be
continuous. This yields an infinite set of linear equations, the solution of which
(after termination) delivers the expansion coefficients. The matching also depends
on the potential in the near-surface region, because it influences the wave function in this region. With the last step we have in principle solved the LEED problem for 'spin-less' particles. While we have briefly sketched the way in which a
proper theory would calculate diffracted intensities, we should like to 'understand', at least in a qualitative way, why the number of peaks in an intensity-
versus energy curve is generally larger than predicted by kinematic theory and why
this latter approach fails. We will follow an argument by Pendry /1974/. The Bloch
states excited in the crystal with wavevector q have a group velocity pointing into the crystal. We imagine that one of these waves (1) approaches another wave (2)
in the band structure diagram with similar q and the same energy, but the opposite
group velocity. Then some part of the flux carried away from the surface by wave
(1) will be carried back by wave (2). This means, that the Bloch state (1) hybrid-
izes with state (2), i.e. it adopts partly the character of a wave with the
'wrong' group velocity. If this happens, the hybridized wave will be excited with
a small amplitude only, and the intensity will be distributed among other beams. A
particularly strong hybridization is observed near cross-overs in the band structure. At these points two Bloch states of equal wave vectors have similar energy
or, conversely, two Bloch states of equal energy are very close in wavevector q.
The parallel components $q_\parallel$ of both states are equal, so that only the normal component has to be considered. Normal components of Bloch states are defined modulo
$2\pi/c_\perp$, where $c_\perp$ is the normal component of the third lattice vector c. The condition for hybridization and with it of enhanced reflectivity then reads

$$q_\perp^{(1)} - q_\perp^{(2)} = \frac{2\pi}{c_\perp} n \qquad n = 0, 1, \ldots \qquad\qquad (2.63)$$

This criterion corresponds to the third Laue condition in kinematic theory. Thus,
the Bragg condition for the wavevector of free electrons is replaced by a Bragg
condition for wavevectors of Bloch states. Because there are many Bloch states excited in the crystal this condition may be fulfilled more often than in the free
electron case. Therefrom the pronounced structure of LEED profiles becomes plausible.

For electrons 'with spin' we have a 2-component free electron spinor in the
vacuum of the form $u_{-k}(\underline{r}) = \exp(i\underline{k}\underline{r})(a_k \cdot |\alpha\rangle + b_k \cdot |\beta\rangle)$ for the primary beam and for
the diffracted beams the analogous form $v_{-k'}(\underline{r})$. Inside the crystal we also have to

use 2-component (at least) Bloch states, where the lattice periodic function $w_g(\underline{r})$ in (2.62) has the form of a spinor: $\underline{w}_q(\underline{r}) = \phi_q(\underline{r}) \cdot |\alpha> + \eta_q(\underline{r}) \cdot |\beta>$. When the amplitudes of the outgoing beams have been obtained in the way sketched above, we know the scattering matrix $\underline{\underline{S}}_{kk'}$. From this we may determine the density matrix ρ' of the diffracted beams, (2.15) from which in turn intensity and polarization may be obtained by means of (2.10) and (2.11).

It should be noted that in principle the intensity has to be determined in this way, even if one is not interested in the polarization at all. The reason is, similar to the case of free particles, that the spin-flip or exchange amplitudes contribute intensity (cf. Sects. 2.2,3).

The spinor field consisting of the primary beam and all relevant ingoing and outgoing spinors matched across the surface is called the 'LEED state'. This state is of great importance not only in LEED, but in all electron spectroscopies at surfaces, for example in electron energy loss spectroscopy, secondary electron emission or photoemission. It will be shown below that the final state in photo-emission is identical to the time reversed LEED state in Fig. 2.9.

Within the framework of the LEED state composed of plane-wave spinors and Bloch spinors, the kinematic theory (considering only plane wave states) follows from the dynamical theory in the limit $\underline{g} \equiv 0$. this means that in the plane wave expansion of (2.62) all Fourier components except the first one are neglected. As mentioned above, the higher order terms contribute the less intensity the larger the imaginary part of the wavevector, i.e. the stronger the damping of the corresponding plane wave spinor. Thus, the kinematic limit is reached for strong absorption, in agreement with our above qualitative discussion of multiple scattering.

Potential Models

A dynamical LEED theory has to treat the propagation conditions for kinetic electrons in a semi-infinite crystal, and to calculate the intensity and spin polarization of electrons leaving the surface. This can be viewed as a scattering problem from a semi-infinite periodic potential. The first part of the task corresponds, except for the damping, to the calculation of the elastic scattering of conduction band electrons in a periodic array of ion cores, i.e. to the band structure. Therefore it is not surprising that some concepts and models of band-structure calculations (see e.g. Callaway /1964/) have been used in LEED theory.

One such model is the 'muffin-tin' model for the periodic potential. This approximation assumes the weakly varying potential between the ion cores to be constant. The scattering potentials are approximated by spherically symmetric potential at the site of the lattice atoms, which are embedded into the constant potential in such a way that the spheres do not overlap. For the tightly bound electron states the free atom orbitals are taken. The free atom valence states usually extend beyond the muffin-tin radius, which requires a renormalization of the wavefunctions, with charge conservation imposed. This model, though limited, has had considerable success in band-structure and LEED calculations.

The interaction of the kinetic electrons with the electrons of the solid is a complicated many-body problem, in principle. It is reduced to a single electron problem by the introduction of a complex 'inner potential': $V_0 = V_{or} + iV_{oi}$. The imaginary part describes the inelastic interaction, while the real part describes the 'elastic' interaction of the primary electrons with those of the crystal. An electron entering a solid experiences a gain of potential energy, i.e. the work function $e\phi$. The maximum energy of conduction band electrons in metals is given by the Fermi energy, relative to the lower edge of the conduction band. If the muffin-tin zero is close to the zero of the conduction band, the real part of the inner potential V_{or} is equal to the sum of work function and Fermi energy. An electron far from the surface induces a charge distribution at the surface such that its field outside the surface corresponds to that of an image charge inside the metal. The resulting 'image potential' varies asymptotically as $\sim 1/4z$. In the 'jellium model' the positive potential of the ion cores is smeared out uniformly, preserving charge neutrality, while the electron gas is of uniform density, except at the surface, where a certain 'spill-over' into vacuum occurs. If the electron approaches such a jellium surface, the image force does not diverge. Rather, there is a continuous transition from the two-dimensional charge distribution that generates the Coulomb field, to the spherical distribution of the 'exchange-correlation hole' inside the solid. The exchange-correlation energy μ_{xc} is gained by the electron in addition to the energy $e\phi_D$ it gains from passing through the surface dipole layer, which is generated by the spill-over of the electron charge density. Within the jellium model we have approximately for the real part of the inner potential

$$V_{or} \approx \mu_{xc} + e\phi_D = E_F + e\phi \quad . \tag{2.64}$$

In a better approximation the inner potential has to be assumed energy dependent, as the assumption of adiabatic processes, which we made implicitly, is not necessarily valid. Also, the one-dimensional jellium model does not give an exact description of the three-dimensional potential the electron experiences when approaching the surface. It turned out, however, that a simplified shape, such as sketched in Fig. 2.10 gives rather good results, provided the reflection and transmission properties of the surface barrier are 'right'.

As far as the spin polarization is concerned, it is important to note that the surface barrier itself does not give rise to polarization /Feder and Meister, 1969/. The reason is that the potential gradients are far too weak to cause noticeable spin-orbit interaction. (For an infinite potential step this would not necessarily be the case, Weisskopf /1935/). However, the surface barrier may have

34

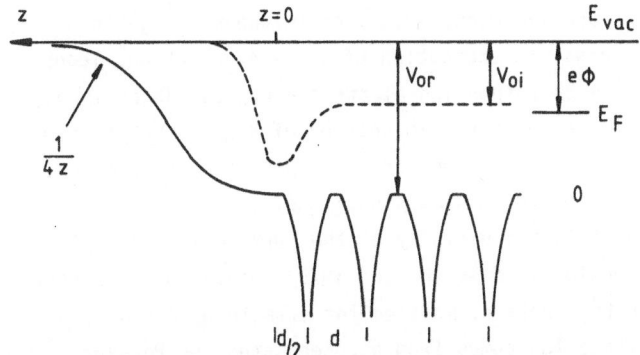

Fig. 2.10. Schematic representation of the real and imaginary part of a model potential such as used in dynamical LEED calculations for W. The real part is characterized by the 'muffin-tin spheres', the asymptotic 'image potential' ($\sim 1/4z$) and the smooth transition (according to the 'jellium model') to the muffin-tin zero. The imaginary part V_{oi} is assumed constant inside the bulk, with a localized part at the surface. For tungsten d = 0.158 nm

considerable indirect influence due to its transmissive and reflective properties (see Chap. 4).

We may determine the approximate value of the real part of the inner potential from the electron theory of metals (2.64). More precise values can be obtained by comparing experimental and theoretical LEED I-V curves. The inner potential causes a reduction of the wavelength of the electrons inside the crystal. For normal incidence and normal reflection ($\underline{k}_\parallel = \underline{k}'_\parallel = \underline{g}_\parallel \equiv 0$) a change of V_{or} only changes $g_\perp$, which means that the theoretical I-V curve of the (0,0) beam remains the same except for a shift on the energy scale. This is approximately true also for other low index beams. By fitting intensity maxima of calculated and measured curves on the energy axis V_{or} may be determined as a function of energy. It turns out that V_{or} is only weakly energy dependent, with a typical value around 10 eV, which agrees with results from (2.64).

The imaginary part of the inner potential V_{oi} describes the loss of electrons from the elastic channel due to inelastic processes. It may be estimated from experimentally known values for the mean free path /Seah and Dench, 1979/. The time-dependent wavefunction of electrons (no spin) is of the form

$$\psi(\underline{r},t) = \psi(\underline{r})\, e^{-i\frac{E}{\hbar}t} \, . \tag{2.65}$$

With the inner potential iV_{oi} the intensity decays in time with the factor $\exp(2V_{oi}t/\hbar)$. Equating $2V_{oi} = -\frac{1}{\tau}$ we connect V_{oi} with the mean lifetime τ, which is related to the mean free path λ at electron velocity v_e by $\tau = \lambda/v_e$. For typical electron energies used in LEED one finds values around $V_{oi} = 4$ eV, which corresponds to $\lambda = 0.6$ nm. To assume the imaginary part spatially constant is an approximation that may or may not be justified, depending on its success /Feder, 1981/. In the case of W, for example, it was noted by Feder and Kirschner /1981a/

that introducing an additional absorptive part, localized in the muffing-tin spheres to take into account the possible excitation of the W 4f electron, leads to a significant improvement at the corresponding electron energies. There is also a study by Rasolt and Davis /1979/ in which the anisotropy of V_{io} was taken into account. The level of improvement is of the same order as when using a more real-istic surface barrier model instead of the 1-dimensional one.

The shape of the absorptive part in the vicinity of the surface is also not well known. High resolution reflection studies /McRae, 1979/ showed that it should comprise a short range part near the surface, modeled for example by a Gaussian like in Fig. 2.10. Further justification comes from a recent study by Persson /1983/ on inelastic scattering, that showed a large surface related contribution from electron-hole pair excitations.

In the case of magnetic materials the inelastic mean free path as well as lo-calized absorption may be spin-dependent. This is discussed in Sect. 2.6.

Calculations

With the above or similar models a number of calculations have been carried out by Feder /1971,1972,1974,1977a/, Jennings and Jones /1978/ and Jennings /1970, 1971a,b/. A comparison between models has been made by Feder, Jennings, and Jones /1976/. A detailed description of the theoretical procedures may be found in the articles by Pendry /1976/ and Feder /1981/. Here, we will only briefly explain the principle of the calculations.

As a solution to the multiple scattering problem with spin-orbit coupling a re-flection matrix $\underline{\underline{R}}$ is wanted, that describes the backdiffracted wavefield as a function of energy, polarization and orientation of the primary beam $\underline{u}_k$. This ma-trix contains for each diffracted beam $\underline{v}_{k'}$ the corresponding spin-scattering ma-trix $\underline{\underline{S}}_{kk'}$, which connects $\underline{v}_{k'}$ with $\underline{u}_k$. The solution is found in several consecu-tive steps. First, the crystal is decomposed into layers along z, that contain one or more crystallographic planes of the crystal. The surface region is, depending on the actual problem, represented by one or more further layers.

The scattering from a single muffin-tin sphere is described by calculating the spin- and energy-dependent scattering phase shifts (cf. (2.30)). Here enters the kind of atoms forming the crystal. The 'small' components of the four-spinor are not necessarily small in this step, which requires the full Dirac equation to be used.

The temperature dependent displacements of the atoms are taken into account by replacing the original (real) scattering phase shifts by complex ones. As in the kinematic model the displacements are assumed isotropic and harmonic and are aver-aged over a Debye spectrum with characteristic temperature Θ_D, which may or may not be equal to the bulk Debye temperature. Complex scattering phase shifts atten-uate the scattering amplitudes, thus providing a reduction of multiple scattering.

The multiple scattering problem is further subdivided: multiple scattering within one layer (intra-layer scattering) and between the layers (inter-layer scattering). As the potential is assumed to be constant between the muffin-tin spheres, for these calculations the wavefunctions can be taken as two-component spinors, neglecting the two 'small' components. At each side of a layer a wave-field of N incident plane waves is expanded into spherical harmonics using Clebsch-Gordon coefficients. The incident spherical harmonics generate scattered waves according to the phase shifts calculated above. The total wavefield within a layer is composed of all scattered waves from all atoms of the layer, including multiply scattered waves. The sum can be reduced to a sum over a unit cell due to Bloch's theorem. This wavefield in angular momentum representation is expanded in plane waves and superimposed to the incident wavefield of plane waves in order to obtain the plane waves outgoing from each side of the layer.

The relation between the 2N ingoing and 2N outgoing beams is given by a (4N x 4N) scattering matrix $\underline{M}$ (in the spin-less case only a (2N x 2N) matrix). As the solid consists of a series of identical layers, the multiple scattering be-tween the layers (inter-layer scattering) is known, if there is a 'transfer ma-trix' $\underline{Q}$, that connects the ingoing and outgoing plane waves at layer n with those at layer n+1. As shown by Pendry /1976/ there is an algebraic relation between $\underline{Q}$ and $\underline{M}$, so that $\underline{Q}$ may be obtained from $\underline{M}$. Now each layer is characterized by a transfer matrix and the multiple scattering problem is solved in principle: The eigenvalues of the transfer matrix $\underline{Q}$ yield the band structure and the eigenfunc-tions yield the Bloch-states, expressed in plane-wave spinors. The eigenvalue problem is solved by diagonalizing the $\underline{Q}$ matrix. The wanted reflection matrix $\underline{R}$ is obtained by combining $\underline{Q}$ with another matrix describing the transmission and re-flection of the beams at the surface barrier, and, possibly, with a further trans-fer matrix for an adsorbate layer. The spin-scattering matrices $\underline{S}_{kk'}$ are subma-trices of the reflection matrix $\underline{R}$.

This 'Bloch-wave method' is applicable quite generally, in particular for the case of vanishing absorption, then yielding the ground state band structure. On the other hand, rather large matrices are to be diagonalized (with N = 40 beams a (160x160) matrix), which becomes problematic at high energies since the number of beams increases rapidly. The Bloch wave method is thus most useful at low ener-gies, which means small damping and few beams. For higher energies, a faster meth-od was developed /Feder, 1972/, that assumes a finite crystal. Under this assump-tion the total transfer matrix is given by the product of the transfer matrices of the individual layers times the surface layer transfer matrix. If the plane waves are sufficiently damped, the backscattering from the rest of the crystal may be neglected after a certain number of layers. The amplitudes of the backdiffracted beams then are obtained from the total transfer matrix by zeroing the waves out-going from the bulk. As matrix multiplication is much faster than matrix inversion, and as typically 8 layers are sufficient at LEED energies this method is much more efficient. However, there may arise numerical problems with too many beams and in the case of small absorption, as the crystal then becomes 'too thick'. Therefore for e.g. photoemission or secondary electron emission the Bloch-wave method is preferable, notably after having been improved by Jepsen /1981/ and Feder /1981/.

Up to here we assumed the only spin-dependent interaction to be spin-orbit coupling. In real ferromagnetic crystals there is exchange interaction in addition. The problem of using the full Dirac equation with a non-zero vector potential $\underline{A}$ is quite difficult and has only very recently been attacked (Ackermann and Feder /1984/, Feder et al. /1983/). On the other hand, if spin-orbit coupling is neglected or treated in a separate calculation, the exchange scattering can be handled in a relatively simple way. We recall from Sect. 2.3 that the exchange interaction may be associated with the action of an exchange potential or an effective magnetic field in the Pauli equation. Following Feder /1981/ the effective field may be written as $\underline{B}_{eff} = \Delta V_{ex} \cdot \underline{m}$, where $\underline{m}$ is a unit vector in the direction of the majority spin axis, and ΔV_{ex} is the spin-dependent deviation from the averaged exchange potential $\tilde{V}_{ex}$. The total exchange potential is $V^{\uparrow\downarrow} = \tilde{V}_{ex} \pm \Delta V_{ex}$. With the spin-quantization axis along the magnetization unit vector $\underline{m}$, the total exchange part is

$$\tilde{V}_{ex} + \Delta V_{ex} \cdot \sigma_z = \begin{pmatrix} V^{\uparrow}_{ex} & 0 \\ 0 & V^{\downarrow}_{ex} \end{pmatrix} . \tag{2.66}$$

This means that the spinor components are decoupled, and that the scattering of electrons with spins parallel or antiparallel is described by effective potentials $V^{\uparrow}$ and $V^{\downarrow}$ respectively. Therefore the final result may be obtained from two separate intensity calculations with one-component wavefunctions using the effective exchange potentials. The intensity asymmetry is obtained therefrom in the usual way as a normalized difference. Like in pure exchange scattering the individual spins are conserved, neglecting magnon excitation, and the polarization is equal to the asymmetry: $\underline{P}_{ex} = \underline{A}_{ex}$.

This simplified approach has been shown by Tamura et al. /1984/ to be valid for the low-Z ferromagnets Fe and and Ni. The interference between spin-orbit and exchange scattering in general was found to be negligible by comparing the results of the full relativistic calculation with those from the simplified treatment. For heavy forromagnets like Gd this is no more the case, the correction terms in (2.60) become noticeable and the intensity asymmetry A_u for an unpolarized primary beam upon reversing the magnetization was calculated to reach values up to 40 %.

2.5 Photoemission

The external photoelectric effect, i.e. the emission of electrons from solids by light, was first observed by Heinrich Hertz /1887/. Ironically, this discovery, unexplainable by classical physics, was made by a scientist who considered physics completed with the formulation of Maxwell's equations. The explanation of this effect was given by Einstein /1905/, pointing out the quantum character of the emission and absorption of light. Within the last decade, photoemission has become an indispensable tool in solid state physics in general and surface physics in particular. With the beginning of angular resolved emission studies (Gerhardt and Dietz /1971/, Feuerbacher and Fitton /1973/, Smith and Traum /1973/) and the use of synchrotron radiation /Himpsel and Steinmann, 1975/ the field expanded rapidly. It includes the determination of two- and three-dimensional energy-band dispersions for bulk crystals, surfaces and adsorbates, the identification of surface atoms and -molecules, the orientation and location of atoms and molecules on surfaces. All these investigations rely on the analysis of momentum and/or energy of the photoelectrons, eventually including symmetry relations and light polarization effects. Among the now vast literature the books by Feuerbacher et al. (eds.) /1978/, Kunz (ed.) /1979/, Cardona and Ley (eds.) /1978/ are suggested for further reading, as well as the recent review articles by Plummer and Eberhardt /1982/ and Himpsel /1983/.

Relatively little attention was attributed to the spin of the electrons, which presumably was caused by the traditional difficulties of the spin analysis. Early work was done by Busch et al. /1969/. In these investigations the spin polarization of the total photoelectron yield, mainly from magnetic materials, was investigated to study bulk and surface magnetism. This work is covered in review articles by Alvarado et al. /1978/, Alvarado /1979/ and Siegmann et al. /1984/. An even smaller number of studies was concerned with spin-polarized photoemission from non-magnetic solids. The first experiment was made by Heinzmann et al. /1972/ with polycrystalline Cs, irradiated by circular polarized light. The qualitative explanation was given by Koyama and Merz /1975/ and Koyama /1975/, who pointed out, that within a three-step model of photoemission, the spin polarization is due to the action of the dipole selection rules for circular polarized light in direct interband transitions between occupied and empty bands of the crystal. Much larger polarizations were observed with GaAs by Pierce and Meier /1976/ using single crystals, which laid the basis for a very important class of polarized electron sources (see Sect. 3.2).

In atomic physics the emission of polarized electrons from unpolarized alkali atoms with circularly polarized light was predicted by Fano /1969/ and verified by Heinzmann et al. /1970/. The effect is due to the dependence of the matrix element for dipole transitions on the spin-orbit coupling in the final state. As the electron spin is aligned with the photon momentum, angular resolution is not necessary.

Energy selection is provided by the choice of the photon energy. In 1973/74 it was predicted independently by Cherepkov /1974/ and Lee /1974/ that spin polarization effects should be observed in photoionization of unpolarized atoms by unpolarized light, provided that energy and angular resolution is achieved. These predictions were veryfied in a series of demanding experiments by Heinzmann et al. /1979/, and Schönhense /1980/. In this case also, the effect is due to spin-orbit coupling. For the simplest case its existence may be made plausible by a hypothetical scattering experiment with photoelectrons. We know that in the scattering experiments of electrons from atoms a spin-flip amplitude appears, that is due to spin-orbit coupling while the electron is close to the nucleus. This amplitude causes polarized electrons to appear in certain directions of space. By analogy, such an effect could also arise for photoelectrons, i.e. electrons emitted from the outer electron shells along the electric vector. Therefore polarized electrons may be observed for certain energies and emission angles. It should be noted that in this case, unlike in the Fano effect, the total polarization vanishes if the electron flux is integrated over all angles, and the remaining ions are unpolarized.

We shall see some analogies between spin-polarized photoemission from atoms and solids in the following. First, however, we shall treat the photoemission from solids for the 'spin-less' case, and introduce the spin later.

2.5.1 Photoemission from Crystals

In the previous chapter we recognized LEED as a many-body problem, which could be treated within a single particle model by means of a number of more or less drastic simplifications. The many body aspect is even more pronounced in photoemission. While in LEED the crystal could be assumed to be near its ground state and the inelastic effects could be summarized by a reduction of the elastic particle flux, in photoemission it is inevitably an excited state of the system that is observed. The interaction of the excited electron with the hole created and of the other electrons with that hole cannot be neglected. For example one may think of excitons in insulators and semiconductors, which cannot be treated within a one electron model. In metals no bound electron-hole pairs exist because of the high dielectric constant, but the apparent binding energy of an excited electron is reduced: The electrons of the solid screen the positive hole, the total energy of the system is diminished and because of energy conservation the emitted electron gains a higher kinetic energy ('relaxation'). This effect is well observable for core levels, but plays also a role in emission from the conduction bands, leading to an apparent narrowing of the band width.

The hole has a finite lifetime as it may be filled by electrons in higher energy levels, eventually in cascade processes involving Auger electron emission. The fact that the Fermi edge is 'sharp' in a photoelectron spectrum follows from the phase space argument that only few electrons are available to fill the hole.

Therefore the lifetime increases and the energy uncertainty decreases. Typical broadenings in nearly free electron metals (Al) are around a few tenth of an eV at 1 eV binding energy, increasing to several eV at 10 eV binding energy. Likewise, the lifetime broadening of the photoelectron ranges from ~1 eV at 10 eV above E_F up to ~10 eV at 100 eV kinetic energy /Levinson et al., 1983/.

Implicitly we have assumed the excitation process to be adiabatic i.e. suffi-ciently slow for the electronic system to relax completely. This assumption needs not be fulfilled (in fact it never is completely), so that other excitations may occur simultaneously with the electron emission, for example the excitation of volume and/or surface plasmons or of low energy electron-hole pairs /Gadzuk, 1978/. The emitted electron lacks the energy consumed in these processes, so that asymmetrical lines with satellites appear. A well known example is a 6 eV energy loss in Ni, which has been observed both in valence-band emission and core-level emission. In both cases, the emission of photoelectrons is accompanied by the ex-citation of a d-band electron into the empty part of the d-band. Since the inter-action with the originally created hole (d-band or core-level) with this addition-al band hole is rather strong, a discrete excited state is observed at higher binding energy /Liebsch, 1981/. These 'intrinsic' losses have to be distinguished from the 'extrinsic' losses which the electron suffers on its way to the detector /Steiner et al., 1979/. Other effects like emission or absorption of phonons (Pen-dry /1976/, Caroli et al. /1978/) and/or magnons during photoexcitation may inter-vene. In view of these complications it is not surprising that a fully general theory of photoemission does not yet exist. There are, however, more or less so-phisticated elements of such a theory, which emphasize the particular aspects relevant for a given experiment.

A substantial simplification is achieved if a one-electron model can be used, either if the above many-body aspects are small, or by summarizing them in cor-rections applied to the one-electron model (Davis and Feldkamp /1980/, Tréglia et al. /1980/, Sacchetti /1980/). That this is viable in many cases has been proven by the success of the band mapping techniques /Himpsel, 1983/. If we further dis-regard the existence of the crystal surface for the moment, assume three-particle events (e.g. electron-phonon-photon) to be unlikely, and the electron and hole lifetimes to be sufficiently long, we arrive at the following simplified picture of the electron-photon interaction: The absorption of a photon causes a transition between the occupied part and the empty part of the (ground-state) band structure, while energy and quasimomenta are conserved. For common photon energies of 1-1000 eV the momentum of the photon is negligible relative to typical electron wavevec-tors $\underline{q}$. In this approximation the wavevector $\underline{q}$ is conserved modulo $\underline{g}$. In the re-duced zone scheme, the photon of energy $\hbar\omega$ causes a direct transition across the Fermi-level within the *same* band structure for the initial and the final state. In the following we will discuss the nature of the electron-photon interaction, the transition probability and the initial and final states in some more detail.

Generally, the transition probability is calculated within the semiclassical dipole approximation, which starts from the non-relativistic Schrödinger equation.
Though this ansatz may be questionable in principle, in particular with respect to
spin-polarized photoemission, it is useful as a starting point. A fully relativ-
stic theory is currently being developed by Feder, Borstel and coworkers.

In the semiclassical approximation the fields are treated classically while the
particles in the field are treated quantum mechanically. This procedure is justified for small photon energies, as the fully quantum mechanical calculation gives
the same results /Schiff, 1955/. The electron spin is not explicitly taken into
account, though implicitly by the symmetries of the band structure and the selection rules. Under these conditions in the Hamilton operator of (2.20) all terms
containing the spin operator $\underline{\sigma}$ are neglected and we end up with the non-relativistic Schrödinger equation for a charged particle in an external electromagnetic
field. The photon flux densities are small, which allows to neglect nonlinear effects and to treat the field as a perturbation. For the perturbation operator we
obtain from (2.20), neglecting terms in $\underline{A}^2$, an expression of the form

$$H' \sim (\underline{A}\nabla + \nabla\underline{A}) \quad . \tag{2.67}$$

First order perturbation theory ('Fermi's golden rule') gives the transition probability per unit time W_{fi}:

$$W_{fi} \sim |<\psi_f|\,\underline{A}\nabla + \nabla\underline{A}\,|\psi_i>|^2 \cdot \delta(E_f - E_i - \hbar\omega) \quad . \tag{2.68}$$

$\underline{A}$ means the vector potential in the crystal. The term $\mathrm{div}\underline{A}$ can be made zero in the
vacuum by a gauge transformation. At the surface, however, mobile charge is avail-
able, the response of which has to be considered explicitly (Forstmann and
Stenschke /1977/, Ashcroft /1978/). Within the jellium model it was shown bei
Feibelman /1975,1976/ that the component of $\underline{A}$ normal to the surface shows rapid
spatial oscillations that are comparable to typical lattice constants. It was also
shown experimentally that these may have considerable impact on the photoelectric
yield /Levinson et al., 1979/. Therefore, the assumption of the dipole approximation, that the field is constant over typical dimensions of the system, is not
rigorously fulfilled. The rapidly oscillating field near the surface may be understood as the response of the microscopic system to the classical discontinuity of
the normal component of the $\underline{E}$ vector at an interface. Classically, the parallel
component is continuous, so that one might expect the deviation in the microscopic
picture not to be too strong (a general microscopic theory is missing yet). This
problem should therefore be solved or at least strongly relaxed, if s-polarized
light is used or if the light impinges normally onto the surface.

Even if the $\mathrm{div}\underline{A}$ term is totally ignored, in a classical treatment the polarization state of the light may change: because of the complex index of refraction of

metals the Fresnel formula generally predicts elliptically polarized transmitted
light with linearly polarized incident light. As the polarization state enters the
dipole selection rules, it needs to be known in the crystal. If it cannot be calculated, the problem can only be solved by normal incidence of the light.

In the following let us make the simplifying assumptions that the vector potential be known inside the crystal and that the rapid oscillation of $\underline{A}$ near the surface be negligible. Then the term div$\underline{A}$ may be neglected and the matrix element in
(2.68) reads

$$M = <\psi_f|\underline{A}\nabla|\psi_i> \quad . \tag{2.69}$$

By means of the commuting relations for the non-relativistic Hamilton operator
$H = p^2/2m + V(r)$ one may write:

$$M = <\psi_f|\underline{A}\underline{p}|\psi_i> = \frac{1}{\omega} <\psi_f|\underline{A}\nabla V|\psi_i> = \text{Im} \ (\omega <\psi_f|\underline{A}r|\psi_i>) \quad . \tag{2.70}$$

Implicitly we assumed that both wavefunctions ψ_i and ψ_f are eigenfunctions of the
same operator. In theoretical photoemission calculations generally this simplified
form of the matrix element is evaluated.

Because of the equivalence of the representations the most suitable one for the
particular application may be chosen. If, for example, the muffin-tin model is
used, one will choose the grad V representation. In the space between the muffin-
tin spheres there are no transitions ($\nabla V \equiv 0$), and it is sufficient to calculate
transitions within the spheres, using a spherical harmonics representation of the
wavefunctions.

Initial and Final States

Let us first consider a relatively simple case. The photon energy is large
($\sim 10^3$ eV) and the initial state is a conduction-band state. After the excitation
conserving energy and wavevector the kinetic electron is 'far up' in the band
structure and may be regarded as an almost free electron. In the reduced zone
scheme the distance between neighbouring bands becomes very small, so that practically for every value of initial energy and momentum a 'directly' accessible final
state exists /Grobman et al., 1975/. At high energies an experimental integration
over less than 10 % of the half space corresponds to an integration of the whole
Brillouin zone. Therefore the energy distribution gives an image of the density of
states in the occupied part of the band structure. It may, however, be modulated
by dipole selection rules on the matrix element and by the many-body effects discussed above /Steiner et al., 1979/. The mean free path of the electrons is relatively large (typically ~2 nm) so that essentially the bulk density of states is
probed. In combination with spin polarization analysis the spin-resolved density
of states of magnetic materials could be investigated - an experiment yet to be
done. Core levels may also be probed. Due to the dispersionless character of these

states they are sharp, except for life-time broadening, and are routinely used for elemental analysis of solids ('ESCA').

In the opposite case of small photon energies (≤ 10 eV) the excited Bloch states are not far from the Fermi level. The lifetime is also rather large which makes the momentum distribution of the photoelectrons in vacuum to be determined by bulk excitations to a large extent. Now, for a given photon energy only a limited number of direct transitions is possible. If we assume, somewhat idealizing, that the electron momentum is transferred undisturbed into the vacuum, an energy- and angle-resolved measurement would overdetermine the system. We have 4 selection rules (three for the components of q, one for energy conservation) in a system with 4 parameters ($\hbar\omega$, kinetic energy, azimuthal angle ϕ, polar angle θ), a further selection rule may be added if the light is polarized. (A particularly lucid description of the selection rules due to parity conservation has been given by Goldmann /1982/.) At constant angles the energy spectrum ideally would consist of a sequence of δ functions, as well as at constant energy the intensity as a function of one angle. Including the finite lifetime and finite experimental resolution, the spectra would be composed of a number of Gauss- or Lorentz-broadened lines. Indeed, energy spectra have been measured (see e.g. Feuerbacher, Fitton and Willis /1978/) that agree with this picture surprisingly well, even at electron energies of 15 eV above E_F. Limitations to this simple picture are discussed below. The interpretation of such energy spectra has been made largely within the framework of the 'three-step-model', in various degrees of refinement. It was suggested by Fan /1945/, improved by Mayer and Thomas /1957/, and has been much used since then, in particular by Spicer and coworkers (Berglund and Spicer /1964/, Feibelman and Eastman /1974/). To lowest approximation, the intensity is given by the matrix element (2.69), while the finite escape depth of electrons and penetration depth of photons are described by absorption coefficients. The surface is modelled by a transmission coefficient and the refraction at the surface barrier. This model neglects the diffraction of electrons at the surface, which now, in contrast to LEED, approach the surface from the crystal side. The next step of conceptual refinement therefore leads to the semiclassical picture sketched in Fig. 2.11. Electrons in upper Bloch states, characterized by $q^{(1)}$ and $q^{(2)}$ and positive group velocity are partially transmitted through the surface. Because of the 2-dimensional periodicity of the surface the parallel component $q_\parallel$ is conserved modulo $g_\parallel$. The amplitudes of the transmitted plane waves have to be determined from the matching conditions. There is a further complication in that the detector may see intensity from two or more Bloch states that have different $q_\parallel$ inside. The solid beam in Fig. 2.11 has $\underline{k}_\parallel = g_\parallel^{(1)} + (0,0)$, while the parallel dashed beam results from the Bloch state $q^{(2)}$ via the reciprocal lattice vector $g_\parallel$ ('surface umklapp'). Explicit calculations by Jepsen /1979/ for Cu show that the expansion coefficients of these partial waves are about one third as large as the main wave. This results typically in a 10 % contribution for clean surfaces,

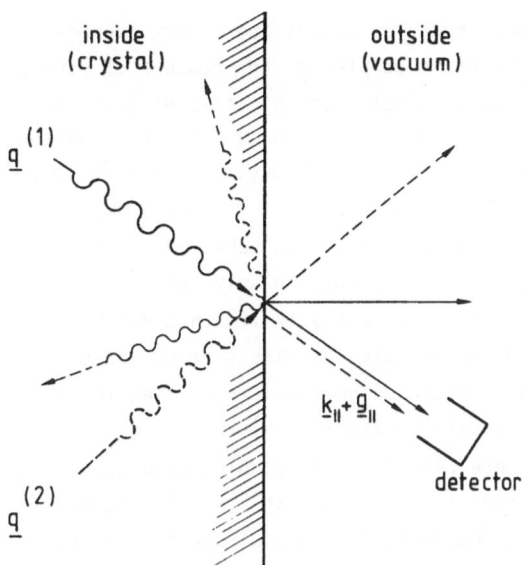

inside
(crystal)

outside
(vacuum)

$\underline{q}^{(1)}$

$\underline{k}_\parallel + \underline{g}_\parallel$

$\underline{q}^{(2)}$

detector

Fig. 2.11. Semiclassical picture of the transmission of photoelectrons through the solid-vacuum interface. Note that this picture does not represent the final state $\langle\psi_f|$ in the matrix element of (2.68). $\langle\psi_f|$ is given by the LEED state of Fig. 2.9, with the detector replacing the source

which may become considerably larger with adsorbate layers /Westphal and Goldman, 1983/. In the case of a reconstructed surface or an ordered adsorbate layer also non-integer $g_\parallel$ may contribute.

The semiclassical wavefield in Fig. 2.11 resembles a 'reverse LEED' experiment with two sources in the crystal. It is emphasized, however, that this state is *not* the correct one to use as the final state in a proper theory. In a proper quantum-mechanical treatment the separation into several steps, like in the three-step-model, is not appropriate. Rather, the matrix element has to contain the directly observed state as the final state, i.e. the wavefunction that describes the intensity at the detector in the vacuum. For orthogonality reasons the outgoing plane wavefield has to be complemented by ingoing plane waves and the final wave function has to be formed from that wavefield. The quantum-mechanical 'one-step theory' (Adawi /1964/, Mahan, /1970/, Schaich and Ashcroft /1971/, Caroli et al. /1973/, Pendry /1976/, Bross /1977/) shows that the bra-vector $\langle\psi_f|$ is exactly given by the 'LEED state' of Fig. 2.9. The reflected Bloch waves in Fig. 2.11 do not play any role. In other words, the final state $|\psi_f\rangle$ in photoemission is identical to the 'time-reversed LEED state' (cf. (2.50)). It is obtained from Fig. 2.9 by replacing the source by a detector and by inverting all pointers indicating the group velocity. This fundamental relation between LEED and photoemission also has consequences for spin polarization effects in photoemission (see Sect. 2.5.2).

As we know, the Bloch components of the LEED state are damped by inelastic processes. They are eigenfunctions of the complex band structure of the crystal. A one-step theory therefore does not describe transitions between the (real) ground-state band structure of the crystal - at least the final state is characterized by a complex wavevector q. (Because $g_\parallel$ is a 'good' quantum number only $q_\perp$ becomes complex). As the initial state $|\psi_i>$, assumed to be an eigenstate of the ground-state, is real in q, and as the final state has a complex q, the concept of direct transitions (i.e. q conservation) becomes questionable. In the limit of bulk effects that we considered up to now, i.e. large mean free path λ, corresponding to a small imaginary part of $q_\perp$, this difference may be negligible. At electron energies around 50 eV, corresponding to mean free paths of a few tenths of a nm, the complex character of the final state band structure has to be taken into account. The analysis of experimental data indicates, however, that in spite of the uncertainty in $q_\perp$ (of the order of $\hbar/2\lambda$, i.e. of 10-20 % of the Brillouin zone) the spectra may still be reasonably well explained in terms of direct transitions /Himpsel, 1983/. It turns out that the final state dispersion often resembles a free electron parabola, eventually with a modified effective mass. This becomes understandable by the fact that band gaps are closed in the complex band structure, which approaches that of free electrons (Nilsson and Dahlbäck /1979/, Hora and Scheffler /1984/) (cf. kinematic approximation). The initial state $|\psi_i>$ is, in principle, also not an eigenstate of the undisturbed crystal, since the remaining hole has a finite lifetime. These effects can approximately be described by a complex q (not just $q_\perp$), where eigenvalues are not necessarily identical with those of the ground state ('lifetime distorted bands'). The band distortions have been found to become particularly pronounced near the zone edges /Nilsson et al., 1980/.

Thus, a proper theory of photoemission describes in a one-step process the transition from an occupied complex band structure into unoccupied, time reversed LEED states and calculates the spatial and momentum distribution in the halfspace in front of the crystal. The level of agreement between theory and experiment is considered quite satisfactory in view of the complexities of the problem (Jepsen /1981/, Westphal et al. /1980/, Hora and Scheffler /1984/), though by no means complete.

There are further specific surface effects we excluded up to now. First, even the ground state band structure of the semi-infinite crystal is complex, due to the existence of surface states ($q_\perp$ imaginary) or surface resonances ($q_\perp$ complex). In bandgaps of the bulk band structure there may exist strongly damped LEED state components, into which direct emission from the surface may occur. Another important area of research is the photoemission from ordered or disordered adsorbates, to gain insight into the electronic and geometric structure of such systems. It is evident, that this subject can only be treated within a one-step theory that describes the transition of the complex band structure of substrate plus adsorbate into LEED states. A theory of this kind (without taking the surface related modifications of the light vector potential into account) has been formulated by Liebsch /1976/. As the LEED state is sensitive to the geometrical structure, intensity analysis allows a determination of the position and orientation of adsorbed atoms or molecules relative to the substrate. This method of 'photoelectron

46

diffraction' (Liebsch /1974/, Scheffler et al. /1978/, Smith et al. /1980/, Kang et al. /1980/, Farrell et al. /1981/) has the particular advantage over LEED that via energy selection certain core levels of the adsorbate can be used. Thus, they act as a primary electron source localized at the surface.

The theoretical concepts discussed so far treat the electron as a 'spin-less' particle. In the following section we will discuss the effects giving rise to a spin polarization of photoelectrons.

2.5.2 Polarized Photoelectrons

We shall base our discussion on the simplified matrix element (2.70)

$$M \sim <\psi_f | \underline{Ar} | \psi_i>$$

In a general case, for example emission from a heavy ferromagnet with elliptical light, the matrix element as a whole will give rise to polarized electrons. For the sake of convenience and simplicity we will discuss each of its ingredients separately, assuming the two other ones to be indifferent with respect to the spin. Under this condition we may distinguish the following cases:

a) *Initial State Effects*. The ground state of a solid may be characterized by a preferential population of electrons with one spin orientation in a band, e.g. in ferromagnets. If the emission from exchange-split bands is energetically re-solved, the spin distribution in a photoelectron spectrum will reflect the spin population of the initial state $|\psi_i>$.

b) *Operator Effects*. In non-magnetic crystals the upper state may become populated preferentially with one spin orientation due to optical selection rules for the dipole operator ($\underline{Ar}$), if spin-orbit coupling is present in at least one of the states involved.

c) *Final State Effects*. If non-magnetic crystals are irradiated with unpolarized light there may polarized electrons be observed above the surface if the final state $<\psi_f|$, which is idential to the LEED state, contains significant spin-or-bit coupling.

a) *Initial State Effects*

The photoelectron yield from ferromagnetic solids has been observed to be partial-ly polarized since some time. The degree of polarization, for example, has been used as a measure for the magnetization of the near-surface layers as a function of temperature (see e.g. Alvarado et al. /1978/, Siegmann et al. /1984/). A more detailed understanding has been achieved since the introduction of momentum analy-sis, which allowed to observe the spin-character of exchange-split energy bands directly /Feder et al., 1983/. The physical concept is fairly simple: if the oper-ator and the final state are insensitive to the electron spin, the degree of po-larization observed outside is directly proportional to the polarization of a particular band. At T = 0 K the bands are expected to be 100 % polarized and so are the emitted electrons. The operator is insensitive to spin if linear or unpolarized light is used. The final state is insensitive to spin in the absence of spin-orbit coupling (and for particular geometries, see below), and if the upper bands are sufficiently high in energy. As the exchange splitting is essentially confined to the d bands in ferromagnets, the energy splitting of the upper bands is small and may be neglected. There are some questions left with respect to spin-flip processes of the excited electrons, i.e. the possible spin dependence of the inelastic mean free path. These questions are discussed in Sect. 2.6. A few examples of 'initial state effects' are given in Chap. 6.

An interesting alternative to photoemission is its inverse process: the emission of photons when electrons injected into LEED states make a direct transition to lower Bloch states. As the lowest states are at the Fermi energy this technique allows to probe the unoccupied levels near the Fermi level. This effect is nothing else but the well-known X-ray bremsstrahlung /Duane and Hunt, 1915/. In complete analogy to photoemission (photon in - electron out) bremsstrahlung spectroscopy probes the inverse process: electron in - photon out. This well-known technique /Merz and Ulmer, 1966/ has been modified and reviewed by Dose and coworkers (Dose /1983/, Scheidt /1983/) and has recently been extended to spin-polarized primary electrons (Unguris et al. /1982b/, Scheidt et al. /1983/). A schematic sketch of this technique is shown in Fig. 2.12. Polarized electrons are injected into upper Bloch states (better: LEED states), from where they fall down to empty spin-split states near the Fermi level. This transition will only occur if the lower state is

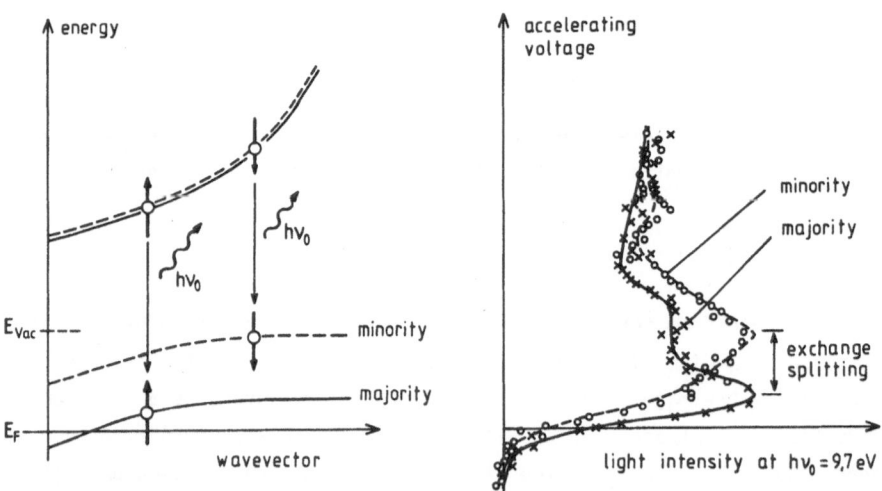

Fig. 2.12. Inverse photoemission in the form of isochromat spectroscopy with spin-polarized primary electrons. The left-hand side represents a schematic bandstructure with spin-dependent direct transitions resulting in the emission of a photon with fixed energy $h\nu_0$. The photon intensity as a function of accelerating voltage is shown on the right-hand side for electrons of opposite spin orientation

of the proper spin character as the spin does not flip during the transition. It is only expected to do so if spin-orbit coupling is active. If the detected photon energy is fixed the technique is more precisely called 'isochromat spectroscopy'. In a scan of photon intensity versus primary electron energy, shown in the right hand part of Fig. 2.12, a peak is observed if for the chosen polarization a band separation of hv_0 is reached. For the opposite spin direction the peak appears at a different electron energy, the energy difference being close to the exchange splitting of the empty bands involved. In addition to the main peaks in the original spectrum there are other transitions visible as well as a rising background due to electron-hole pair excitations. Choosing various angles of incidence the dispersion of empty bands can be mapped out for minority and majority bands separately. A few examples are shown in Chap. 6.

Relative to photoemission the initial and final states are inverted in this technique, but the origin of the polarization effects is the same, namely the different spin character of exchange-split bands.

b) *Operator Effects*

In the 'spin-less' case the perturbation operator (2.67) was obtained from the relativistic Hamiltonian (2.20) by neglecting all terms containing the spin operator σ. This is questionable even if only intensities are considered, as in analogy to electron-atom scattering the intensity might be influenced by spin-flip processes. Calculating intensities only, this may be of minor importance in view of many other simplifications and uncertainties. If the electron spin itself is the quantity of interest, these terms cannot be neglected a priori. The spin-orbit coupling as well as the interaction of the magnetic moment of the electron with the magnetic part of the electromagnetic wave might induce a spin-flip in photon absorption. These influences have been treated by Feuchtwang et al. /1978/ for direct transitions between (ground state) Bloch states. Numerical results have not been given, but it was shown that the ratio of the transition amplitude with spin conservation to that with spin-flip is of the order of $mc^2/\hbar\omega$. For photon energies in the range of $10^1 \ldots 10^3$ eV these effects should therefore be negligible. Thus, for the present state of knowledge the non-relativistic dipole approximation seems appropriate, though a question mark should be kept in mind.

In this approximation spin polarization effects are due only to the optical selection rules for the 'magnetic' quantum number $\Delta m = \pm 1$. We note, that the term 'spin transfer from the photon to the electron', which is sometimes used in the literature in this context, is grossly misleading. It is via spin-orbit coupling only, either in the initial or the final state, that electrons with a certain spin orientation are preferentially excited. In a non-magnetic crystal all bands are doubly degenerate. The Bloch spinors can not only be expanded in plane wave spinors but also into spherical harmonics (in general, not only within muffin-tin spheres). The elements of the expansion consist of a radial function, an angular function (the spherical harmonics $Y_\ell^m(\Theta,\phi)$), and a spin function ($|\alpha\rangle$ or $|\beta\rangle$). The bands of the band structure are characterized by the irreducible representations $\{n\}$ of the point group describing the symmetry at a particular point in the Brillouin zone. There are simple basis functions that obey the symmetry operations of the irreducible representation $\{n\}$. The corresponding wavefunction must have the same symmetry properties as the Hamilton operator. Therefore it can be obtained from a basis function, or a set of orthogonal basis functions in the case of degeneracy. The basis functions can be expanded into the 'symmetry adapted basis functions' $g_n(\Theta,\phi) = g_n(Y_\ell^m(\Theta,\phi))$ by means of spherical harmonics. A number of basis functions was given by Borstel and Wöhlecke /1981/ and by Wöhlecke and Borstel /1981a,b,c,d/ for the most common lattice types. The operator $\underline{A}\underline{r}$ in the matrix element can be reduced to the form $(x \pm iy)$ for circular polarized light propagating along the z axis and considering only electric dipole transitions. As this form corresponds to the spherical harmonic $Y_1^{\pm 1}$ we now see why it is convenient to expand Bloch states into symmetry adapted basis functions. The matrix element consists of a radial part times a product of spherical harmonics. The optical selection rules then are given by the orthogonality properties of the spherical harmonics.

Let us illustrate this point with an example. Consider the Δ axis in a bcc crystal (the [100] axis) and a direct transition at some point on this axis between the bands $\phi_i = \Delta_6^1$ and $\phi_f = \Delta_6^5$ by right circular light. In the convention of classical optics an observer at the crystal looking back at the light source sees the electric vector turning clockwise for right circular light. The helicity is negative in this case as the spin and the momentum of the photon are antiparallel. The operator has the form Y_1^{-1} and the matrix elements are different from zero only if the lower index changes by ± 1 ($\Delta \ell = \pm 1$) and the upper by -1 ($\Delta m = -1$) in the transition:

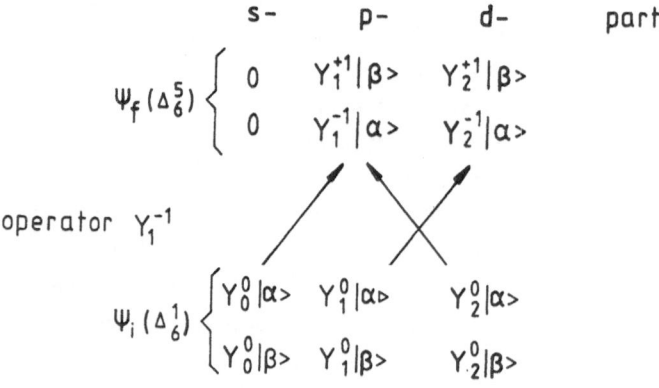

We omitted the radial function in the representation of the basis functions. It is apparent that for the allowed transitions all excited electrons have the spin state $|\alpha>$ for right circular light, i.e. we have 100 % polarized electrons in the upper Bloch state. We neglected transitions with spin-flip $|\alpha> \rightarrow |\beta>$ according to the non-relativistic approximation. If σ^- light was used, the polarization would be reversed but the absolute value would be the same. It should be emphasized that these results depend by no means on the angular momentum representation of the Bloch states. Any other set of suitable basis functions would lead to the same re-sult, as the spin polarization is a consequence of the *symmetry* of the Bloch states, which in turn is governed by the symmetry of the crystal.

We see that strong polarization effects should be observed using circular po-larized light, with the only prerequisite that spin-orbit splitting is present in the upper or lower state and that it is resolved experimentally. Conversely, a measurement of the spin polarization of photoelectrons allows to determine the symmetry character of the bands involved. There is even the possibility to experi-mentally determine the hybridization between bands (Allenspach et al. /1983/, Meier and Pescia /1984/). Hybridization is expected to be strong at points in $\underline{q}$-space where bands of the same symmetry cross or are close in energy. This will be the case when in a band structure without spin-orbit interaction bands were cross-ing, and if the degeneracy is removed by spin-orbit coupling. In this case initial and/or final states are given by a linear combination of symmetry adapted basis functions of the same symmetry. In general one would expect the polarization to be reduced by hybridization, as transitions into opposite spin states may be allowed, but the contrary may be found as well. For example, on the Σ-axis group theory predicts zero polarization for pure states, while hybridization may cause a finite polarization /Borstel and Wöhlecke, 1982/.

There is an important consequence of the reversal of the polarization when re-versing the handedness of the light. Linear polarized or unpolarized light can be thought as a coherent or incoherent superposition of two oppositely polarized cir-cular waves. If follows therefrom qualitatively (though in agreement with the exact results /Koyama, 1975; Borstel and Wöhlecke 1982/ that there are *no* spin-polarized electrons emitted by unpolarized or linear polarized light as far as operator effects are concerned. For a more detailed mathematical discussion of operator effects reference is made to the excellent review articles by Wöhlecke and Borstel /1984/ and Meier and Pescia /1984/.

c) *Final State Effects*

If initial state effects are absent, photoelectrons excited by linear or unpolar-ized light should be unpolarized. This is, however, only true if spin-orbit inter-action in the LEED state $<\psi_f|$ is negligible. It is never absent, though, and as a consequence the electrons observed in vacuum are in general spin-polarized. The total yield still shows zero polarization, but when the photoelectrons are select-ed with respect to momentum, in general a finite spin polarization will be found.

This is somewhat similar to the photoionization of free atoms with linear polarized or unpolarized light which we mentioned above (Sect. 2.5). For crystals, this
phenomenon can be understood by analogy to spin-polarized LEED.

Assume an unpolarized primary beam impinging onto the crystal surface, containing equal numbers of oppositely polarized electrons. The primary beam excites
Bloch spinors in the solid that remove flux from the surface. In the vacuum there
are diffracted beams, with the amplitudes determined by the matching conditions at
the surface. The propagation of Bloch states is spin-dependent due to spin-orbit
coupling in the solid. Therefore the population of the Bloch states becomes non-
symmetrical with respect to the electron spin, and as a consequence of the overall
spin conservation the back-diffracted plane wave spinors are also spin-polarized.
A detector inside the crystal would measure polarized Bloch states. The time reversal theorem allows us to interchange source and detector, which does not change
anything except for the spin reversal. In order to observe zero polarization at
the detector, which is now in the vacuum, we would need a 'Maxwell demon' sending
polarized Bloch spinors and polarized free electron spinors with the correct amplitudes and phases such that they recombine to the previous unpolarized beam at
the detector. If this does not happen, e.g. if the Bloch states are unpolarized,
we will measure a finite polarization of the electrons in the vacuum. The existence of the higher order diffracted beams in the vacuum is not essential in this
argument. The effect would not be altered qualitatively if the energy were so low
in LEED that not backdiffracted beams exist except the (0,0) beam.

These plausibility arguments may be replaced by an equivalent description that
is closer to the quantitative theoretical treatment /Feder and Kirschner, 1981b/.
We make the simplifying assumption that only one excited Bloch spinor q with amplitude 1 approaches the surface. At the surface we observe a transmitted beam,
possibly further diffracted outgoing beams and back-reflected Bloch states. Their
propagation is described by the above mentioned spin-dependent transfer matrix
(Sect. 2.4.3). Its construction according to the Bloch theorem guarantees that an
unpolarized Bloch state propagates as such through the crystal, if the latter has
a center of inversion. According to the Bloch wave method sketched above, Bloch
states in the plane wave representation with N reciprocal lattice vectors are determined by diagonalization of the (4Nx4N) transfer matrix $\underline{Q}$. The matching conditions at the surface yield a system of 4N linear equations for N reflected beams
and N transmitted beams with two components each. From these the amplitudes of the
reflected beams may be eliminated and those of the outgoing beams be determined,
in particular those of the beam going to the detector. The polarization is obtained in the usual way from the density matrix of the detector beam. This treatment differs slightly from the one-step model of photoemission insofar as no ingoing plane wave spinors are required. It is, however, equivalent to the one step
model, as shown by Spanjaard et al. /1977/, if there is only one Bloch state excited in the photoemission process, and it is much simpler to calculate. If sev-

eral Bloch states are involved, an incoherent superposition would be obtained by forming the density matrix from a sum of matrices resulting from different ingoing Bloch states. As far as the (limited) experience tells at present, the assumption of only one final state is fulfilled in many instances /Jepsen, 1979/. In this case, or if all transition probabilities are equal, the polarization is independent of the matrix element, in contrast to the intensity. It is evident that the polarization effects essentially are a matter of the final state. If, for example, at fixed energy and angle of the emitted electrons the photon energy is raised such that a transition from a deeper band appears ('constant final state spectroscopy') the polarization will not change if only one final state is involved, even if the matrix element changes. Conversely, a change of the polarization indicates the contribution of several Bloch states.

It should be noted that final state effects vanish for normal take-off, in analogy to normal reflection in spin-polarized LEED, if the crystal has a center of inversion. Also, the polarization averages to zero upon integration over a cone centered about the surface normal. The magnitude of the effect is, however, not simply correlated to the emission angle relative to the surface normal. There are many counterexamples in polarized LEED! We note also that final state effects may give rise to characteristic intensity asymmetries if the photoelectrons are polarized by initial state or operator effects. They may lead to a left-right intensity asymmetry for complementary take-off angles. Also, and perhaps even more important, the polarization vector measured outside is changed in magnitude and orientation from the 'true' one, i.e. the one that we would observe from initial state or operator effects inside the crystal.

An example for final state effects in photoemission from W(001) is discussed in Sect. 4.2.2.

2.6 Inelastic Processes

In this section we consider the energy loss processes of a 'hot' electron during its interaction with the solid. In non-magnetic solids the electron's interaction with phonons /Ibach and Mills, 1982/, plasmons /Pines, 1963/, and electron-hole pairs (Persson /1983/, David et al. /1970/) has been treated as being independent of the spin of the primary electron. This does, of course, not mean that the electron spin is irrelevant for these materials, as spin-dependent processes like e.g. exchange are always present. Just because there are exactly equal numbers of spin-up and spin-down electrons in the non-magnetic solid there does not remain a net imbalance for incident electrons of opposite spin. In the transition of polarized electrons through thin films of magnetic materials a strong depolarization was observed by Meier et al. /1982b/, which is, however, most likely due to elastic exchange scattering. For the *absolute* inelastic scattering cross-section, we remember, spin-dependent processes are of importance, just as found for the spin-flip amplitude in elastic spin-orbit interaction or the exchange amplitude in elastic exchange scattering. We recall that in modern LEED theory an energy-dependent exchange contribution is incorporated into the real part of the optical potential (cf. Sect. 2.4.3). The spin dependence of the imaginary part is discussed below.

2.6.1 Inelastic LEED

In low energy electron diffraction it was recognized quite early (Turnbull and Farnsworth /1938/, Reichertz and Farnsworth /1949/) that the change in electron momentum and the inelastic interaction are decoupled to a large extent. The large-angle scattering from the ion cores is mostly elastic, except for highly localized electron shells of low binding energy /Feder and Kirschner, 1981a/. The dominant loss mechanisms (plasmon and electron-hole-pair excitations) mostly involve small momentum transfers /Daniels et al., 1970/. Therefore energy loss and diffraction are generally treated as being independent. Considering the inelastically back-diffracted electrons, say in the specular beam, for the sake of simplicity we may distinguish between two channels: (i) The primary electron may have been diffracted first with its full energy and then have lost energy ('diffraction-before-loss') or (ii) it may first have suffered an energy loss and then have been elastically diffracted with its remaining energy ('loss-before-diffraction'). As pointed out by Duke and Laramore /1971/ and Laramore and Duke /1971/ these two processes, because of leading to the same final state, should in principle be added coherently, instead of incoherently. The correct treatment, however, showed that interferences are mostly not of major importance /Duke and Landman, 1972/.

A particularly clear-cut example for the interplay between loss-before-diffraction and diffraction-before-loss has recently been found by Rebenstorff et al. /1984a,b/ in inelastic LEED from Ni (see Fig. 2.13). The upper panel shows the elastically scattered intensity in the (0,0) beam from Ni(110) at an angle of incidence $\theta = 75°$. The oscillatory structure is due to an interference process associated with the emergence of the (1,0) beam, in a way which is discussed in detail in Sect. 4.1.3. For the present purpose we remark only that the energetic position of these LEED fine structures depends very sensitively on the momentum of the incident electron. Fig. 2.13 b-d) shows energy loss spectra at various primary energies. The energy scale refers to the kinetic energy of the scattered electrons. The elastic peak corresponds to the energy in the upper panel (a). We clearly see the interference structure from the elastic process reappear in the inelastic spectrum. This occurs at a fixed kinetic energy after scattering, irrespective of the energy loss relative to the primary energy. The structure therefore does not reflect a structure in the energy loss function. The explanation is, that we do observe the loss-before-diffraction events. To undergo the same interference process as in the elastic channel, an electron must have precisely the same kinetic energy and direction when arriving at the surface. Its surplus energy therefore must have been dissipated before the diffraction process [mainly by

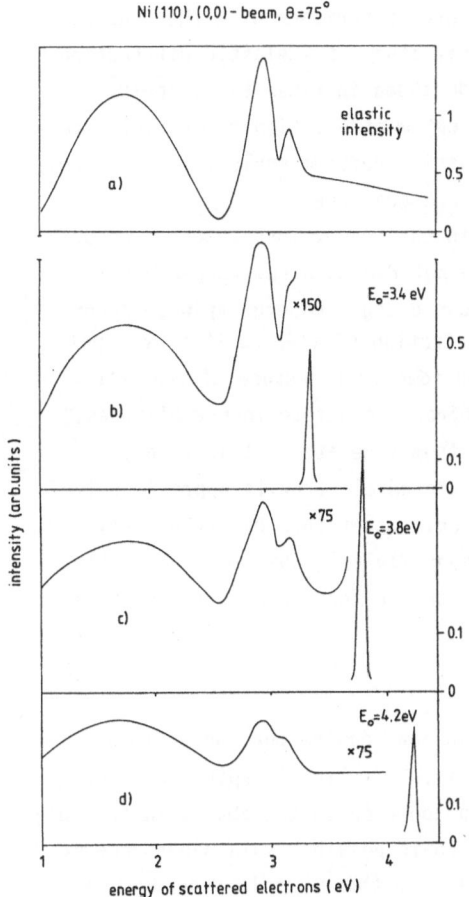

Ni (110), (0,0) – beam, θ=75°

intensity (arb.units)

energy of scattered electrons (eV)

Fig. 2.13. Threshold effects in elastic and inelastic electron diffraction.
a) Elastically scattered intensity in the (0,0) beam as a function of primary energy. b) - d) Energy loss spectra plotted as a function of the energy of the scattered electron for various primary energies E_o

electron-hole pair creation at the surface (Persson /1983/, Sunjic and Penzar /1984/)] without a major change of its momentum direction. But how about the diffraction-before-loss processes? They should be equally probable as the energy losses are fairly small. When the primary energy is above the range of the interference structure, the diffraction process does not introduce sharp intensity variations. The loss spectrum itself is smooth as surface vibrations of adsorbates are absent and the phonon spectrum is not resolved. Therefore the contribution from diffraction-before-loss processes in the inelastic spectrum is a slowly varying function, on top of which the interference structure from loss-before-diffraction processes appears. The latter structure contributes about half of the total loss intensity which means that the two processes contribute by roughly equal amounts.

It may be added that in the elastic channel rather strong spin polarization effects have been found near intensity minima which are due to exchange interaction. Preliminary measurements of the spin polarization in the loss-before-diffraction features show qualitatively similar polarization structure, though at reduced magnitude /Rebenstorff, 1984/. The above findings are at variance with the conclusion of Duke and Landman /1972/ from their theoretical study of inelastic diffraction from Al, that the inelastic spectrum can be understood in kinematical terms. The interference structure is a particularly clear-cut manifestation of multiple scattering and a kinematical treatment of the inelastic spectrum would fail completely as far as the loss-before-diffraction features are concerned.

We note that the spin polarization effects in the energy loss spectrum associated with loss-before-diffraction processes are not due to a spin-dependent loss process. Rather, the electrons, after having lost energy, undergo spin-dependent elastic processes, which result in a spin polarization similar to that of a primary electron of the same energy. (It cannot be identical because of the interference between loss-before-diffraction and diffraction-before-loss amplitudes.) In inelastic scattering of polarized electrons from free atoms it is also frequently observed that the polarization of a scattered electron is approximately determined by its kinetic energy after scattering, rather than by its primary energy (see e.g. Bartschat et al. /1981/ and Hanne /1976/). The same qualitative observation has also been made with amorphous metallic ferromagnets (Siegmann et al. /1981/, Unguris et al. /1984/).

2.6.2 Spin-Dependent Energy Losses

The spin polarization effects discussed above do not require the energy loss process itself to be dependent on the electron spin. Explicitly spin-dependent energy losses may occur independently of and in addition to the above. In the following we shall examine the loss processes and their possible spin dependence in more detail. We focus on ferromagnetic materials, as there a spin dependence may lead to experimentally well observable consequences. Among the explicitly spin-dependent excitations there are spin waves, plasmons, and electron-hole pairs. Spin waves have low energies compared to plasmons, surface spin waves even less /Mills, 1967/, and have not yet been observed experimentally with electrons. Estimates show (Saldana and Helman /1970/, Kleinman /1978/, Feder /1981/) that their contribution to the loss-probability for electrons above 10 eV should be an order of magnitude smaller than that from plasmons and electron-hole pairs. Therefore we confine the discussion to the spin dependence of the latter two elementary excitations in ferromagnets.

A quantity that is sensitive to these loss processes is the inelastic mean free path λ which is related to the imaginary part of the inner potential through (2.65). As we allow it to be spin-dependent, we have to distinguish between $\lambda_{\uparrow}$, which is the mean free path for primary electrons aligned along the spin of the majority electrons, and $\lambda_{\downarrow}$, for electrons with spin parallel to the minority electrons. We may define an asymmetry parameter

$$A_\lambda = \frac{\lambda_\uparrow^{-1} - \lambda_\downarrow^{-1}}{\lambda_\uparrow^{-1} + \lambda_\downarrow^{-1}} \tag{2.71}$$

for the inverse inelastic mean free path which serves as a measure of its spin dependence. Plasmons are collective excitations of the electron gas, involving simultaneously up-spin and down-spin electrons. For a spin-polarized electron gas Helman and Baltensperger /1980/ calculated the inelastic mean free paths for electrons of various energies. From their result it follows that the asymmetry A_λ is rather small, but negative at all energies. This means $\lambda_\uparrow^{-1} < \lambda_\downarrow^{-1}$ or $\lambda_\uparrow > \lambda_\downarrow$.

The same result is qualitatively obtained for the more important electron-hole pair excitations by low energy electrons from the following atomistic argument (Feder /1979/, Bringer et al. /1979/, Feder /1981/). The scattering between a 'hot' electron and a conduction-band electron is thought to be approximately iso-tropic (s-wave scattering) for energies comparable to the Fermi energy. (We notice the difference to the exchange scattering of two free electrons, see Sect. 2.3). Therefore, the exchange amplitude $g(\theta) = f(\pi-\theta)$ is equal to the direct scattering amplitude $f(\theta)$ for all angles, and for electrons of parallel spins the triplet scattering amplitude $f(\theta) - g(\theta)$ vanishes. Somewhat oversimplifying one may conclude that scattering takes place only between electrons of opposite spin orientations. A 'hot' spin-up electron should therefore only be scattered by minority electrons and a spin-down electron only by majority electrons. For a quantitative estimate it was assumed that the ratio of the corresponding imaginary potentials or of the inverse mean free paths is equal to the ratio of the corresponding scattering partners $n^\uparrow$ ($n^\downarrow$) i.e. the number of spin-up (spin-down) valence electrons per atom. It follows for the asymmetry

$$A_\lambda^o = \frac{n_\downarrow - n_\uparrow}{n_\downarrow + n_\uparrow} \quad, \tag{2.72}$$

which is negative as $n_\uparrow > n_\downarrow$ by definition. Clearly, any energy dependence is missing in this 'zeroth-order' estimate. With increasing energy the cancellation of exchange and direct scattering amplitudes should gradually disappear as the scattering involves higher order phase shifts. The asymmetry should therefore decrease with increasing energy, but it should stay negative.

These conclusions were also obtained by Matthew /1982/ for a more detailed atomistic model. The spin dependence of the inelastic mean free path was calculated for polarized electrons ($\gtrsim 100$ eV) scattered from oriented atoms in the Born-Ochkur approximation, a form of the Born-Oppenheimer scattering approximation. In the vicinity of inelastic thresholds the same expression as above was found, while for higher energies E the result was

$$A_\lambda' = \frac{n_\downarrow - n_\uparrow}{n_\downarrow + n_\uparrow} \left(\frac{<\Delta E>}{2E}\right)^2 \tag{2.73}$$

with $\langle\Delta E\rangle$ the mean excitation energy obtained by weighting all possible transition energies by the relevant cross sections. It was estimated to be of the order of 20-25 eV for transition metals. Thus, at $E \gtrsim 100$ eV the asymmetry A_λ' is of the order of -0.3 % for Fe, which may be compared to $A_\lambda^0 \approx -28$ % using the low energy estimate above.

An entirely different point of view was chosen by Rendell and Penn /1981/ by treating the conduction-band electrons, including the d electrons, as an electron gas in a kind of local density approximation. Extending the work of Ritchie and Ashley /1965/ to the case of ferromagnetic materials and using the Born approximation they found similar results as Matthew /1982/ for energies around 100 eV, though the decrease of the asymmetry was not as fast with energy. At energies below about 100 eV, however, Rendell and Penn found a reversal of the sign of the asymmetry, meaning $\lambda_\downarrow > \lambda_\uparrow$. The origin of this discrepancy is not yet clear - it may be due to lumping the s- and d-electron densities together in the electron gas /Matthew, 1982/. The magnitude was found to be of the order of a few percent at low energies.

The theoretical situation obviously is not clear at the present time. It appears likely, however, that the asymmetry of the mean free path is negative at high energies and that its magnitude is of the order of a percent or less. At low energies the theoretical results are contradictory and a clear-cut experimental evidence is still missing. The argument by Bringer et al., however, appears attractive and is qualitatively not in contradiction to experimental results from a ferromagnetic glass /Siegmann et al., 1981/. Quantitatively, however, the effects on the mean free path appear overestimated.

The mean free path is a rather crude measure of the spin dependence of electronic excitations since an average is made over all possible energy losses and their spin dependence. It is likely that for particular loss energies and/or scattering conditions much larger effects will appear. For example, as will be demonstrated in Chap. 5, large asymmetries are observed near excitation thresholds of core levels. In this case the dominant quantitly is the density of empty states available to the primary electron having lost almost all its energy. For these particular processes the excitation probability for down-spin electrons is larger than for up-spin electrons because they have a high empty density of states available.

In the above discussion the process of 'spin-flip' scattering was not considered explicitly. Even in the absence of processes turning around the electron spin (like e.g. spin-orbit interaction and magnon excitations), 'spin-flip' processes may occur if the primary low energy electron falls into one of the empty states above the Fermi level and kicks out an electron of opposite spin from one of the filled bands. Though the final state looks 'as if' the primary electron had flipped its spin during the energy loss, this is not really the case. The electric and magnetic fields involved are far too small. The possibility of these processes

58

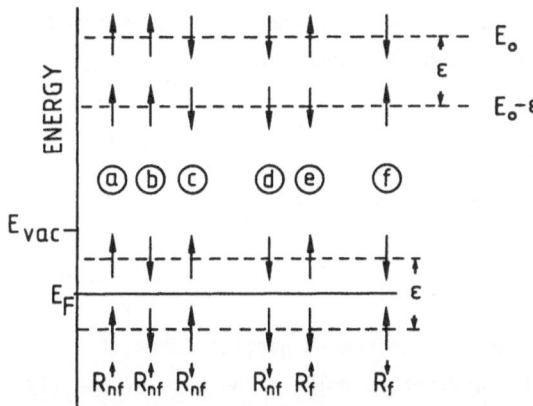

Fig. 2.14. Two-electron processes with spin included as observed in inelastic electron scattering from ferromagnets. Electrons with energy E_0 and a particular spin orientation are sent to the sample, and electrons with energy $E_0-\varepsilon$ are detected by an analyzer. The energy loss ε equals the energetic difference between initially occupied levels below the Fermi energy and the levels above E_F populated after the excitation process

to contribute to the energy loss spectrum has been pointed out by Yin and Tosatti /1981/ and treated in a model calculation for partially occupied exchange-split free electron bands. For purposes of later use we list in Fig. 2.14 schematically the processes that may occur in an experiment where a primary electron with energy E_0 and particular spin orientation is launched onto the crystal and an electron with energy $E_0-\varepsilon$ is observed after scattering. A primary up-spin electron may undergo two loss processes, each resulting in a scattered electron with the same spin orientation as the incident one ('non-flip'). It either excites a majority electron from below the Fermi level to a state above E_F, or it excites a minority electron to the same energy. There are thus two 'up-spin non-flip intensities' $R_{nf}^{\uparrow}$ (a) and (b). The first one has an exchange contribution, which partly cancels the direct amplitude. We note that the non-flip intensities result from excitations within the same spin system [processes a), b), c), d)]. This is essentially the same as is observed in an optical absorption experiment because the electromagnetic fields involved are far too small to reverse the spin of an electron. Within this approximation (i.e. disregarding exchange processes) electron energy loss spectroscopy and optical absorption are equivalent. Differences come about by the possibility of electron exchange [processes e) and f)]. In the 'spin-flip' process described by $R_f^{\uparrow}$, for example, the primary up-spin electron transfers its whole energy to a minority spin electron and occupies an empty state above the Fermi-level after the excitation. The excited minority electron, coming from below the Fermi level, emerges with an energy equal to the primary energy minus the energy of the electron-hole pair around the Fermi-level (Stoner excitation). The partial intensities for a primary down-spin electron are obtained analogously.

Following Yin and Tosatti /1981/ we may define an inelastic scattering asymmetry $A_{in}(E_0,\varepsilon)$ which describes the relative intensity differences at a particular energy loss ε if the polarization of the primary beam with energy E_0 is reversed. Lumping together all spin-flip and non-flip intensities one finds:

$$A_{in}(E_0,\varepsilon) = \frac{(R_{nf}^{\uparrow} + R_f^{\uparrow}) - (R_{nf}^{\downarrow} + R_f^{\downarrow})}{(R_{nf}^{\uparrow} + R_f^{\uparrow}) + (R_{nf}^{\downarrow} + R_f^{\downarrow})} \; . \tag{2.74}$$

Likewise, an initially unpolarized beam may become polarized:

$$P_{in}(E_0,\varepsilon) = \frac{(R_{nf}^{\uparrow} + R_f^{\downarrow}) - (R_{nf}^{\downarrow} + R_f^{\uparrow})}{(R_{nf}^{\uparrow} + R_f^{\downarrow}) + (R_{nf}^{\downarrow} + R_f^{\uparrow})} \; . \tag{2.75}$$

Though the polarization is obtained from the same scattered partial intensities, it is in general not equal to the intensity asymmetry. Only in the case of exactly equal or vanishing spin-flip contributions we have A = P, the familiar result from elastic exchange scattering.

A few experimental results of inelastic q-resolved spin-dependent scattering from Ni will be discussed in Chap. 5. We shall see how the spectrum of Stoner excitations at q = 0 may be probed by spin-polarized electron energy loss spectroscopy.

3. Experimental Considerations

The rapid development of surface physics with spin-polarized electrons owes much to the development of new detectors and sources of polarized electrons. Many new experiments have become feasible only by these developments. In this chapter we will deal with the modern developments of sources and detectors, the traditional instruments being well described in other articles. A lucid explanation why conventional spin filters do not work with electrons has been given by Kessler /1976/. The Mott-Detector has extensively been treated by Van Klinken /1966/, Eckstein /1970/, and Jost et al. /1981/. Sources of polarized electrons, either by spin-orbit scattering from free atoms or by photoionization are described by Kessler /1976/. Sources based on Penning ionization of optically pumped atoms have been developed by Keliher et al. /1975/. The technique of electron capture spectroscopy using ions reflected under grazing incidence from magnetic materials has been reviewed by Rau /1982/.

A certain number of electron spin polarization experiments may be carried out either using a polarized electron source and an intensity measurement, yielding the asymmetry $\underline{A}$, or using an unpolarized source and a spin polarization detector, yielding the polarization $\underline{P}$. We have seen in Chap. 2 that these two approaches are equivalent for elastic scattering experiments as the magnitudes of the polarization and asymmetry vectors are equal: $|\underline{A}| = |\underline{P}|$. The orientations may be different, depending on particular symmetry properties of the system studied, but they are in a fixed relationship. In these cases it is largely a matter of experimental convenience which approach is used. In inelastic scattering this equality does not hold any more as we have seen in Sect. 2.6. A complete determination of the scattering amplitudes requires a polarized primary beam in addition to polarization analysis. In emission experiments like photoemission or field emission a detector is mandatory, anyway.

3.1 Detectors of Polarized Electrons

Electron spectroscopies at surfaces generally are carried out under Ultra-High Vacuum conditions (UHV) to avoid contamination from the residual gas. The two modern detectors discussed in this chapter, the LEED detector and the absorbed current detector both require UHV, but this is in general no obstacle. In atomic physics experiments the vacuum conditions mostly are less stringent, though in advanced electron-atom scattering experiments extreme vacuum conditions are also required. In these circumstances (R. Celotta, private communication) the LEED detector may be used.

3.1.1 The LEED Detector

The operating principle of the LEED detector is the left-right asymmetry due to spin-orbit coupling in electron-atom scattering. This is the same principle as used in the traditional Mott-detector, with the difference that the target is a single crystal surface. In the simplest case, the intensity of the specularly reflected beam from a high-Z material is measured while the crystal is rotated about the incident beam, like in the Davisson-Germer experiment, or while the polarization of the incident beam is reversed. The apparatus is of extreme simplicity, requiring little more than a crystal and a Faraday cup or electron multiplier. Such a device is e.g. useful for monitoring the polarization of an electron source. The intensity asymmetry may be close to 90 % for W(001) near a minimum of the reflectivity (see Chap. 4).

A more convenient design avoiding mechanical manipulations is shown semi-schematically in Fig. 3.1. In this set-up the primary beam impinges normally onto a W(001) surface. Among the diffracted beams the ($\bar{2}$,0) and (2,0) beams are detected by two channeltrons looking at the crystal under a fixed angle corresponding to the exit angle of the beams for a certain scattering energy. The inelastic background is removed by two simple retarding fields in front of the channeltrons. At this stage several other detection schemes could be used that have been developed for conventional LEED work /Lagally and Martin, 1983/. The energy resolution needs not to be very high, an energy width of 2 to 5 eV can be tolerated without severe loss of polarization sensitivity. For unpolarized normally incident electrons the four (2,0) beams are degenerate, i.e. are of equal intensity. If the incident beam has a polarization vector normal to the electron momentum and normal to the paper

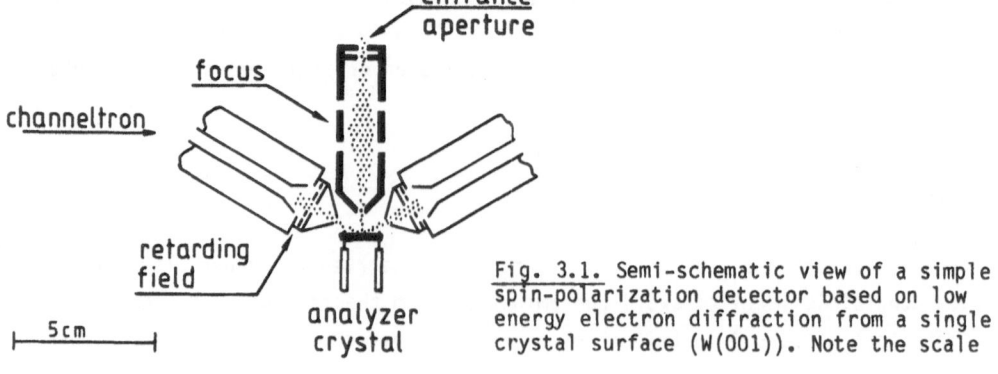

Fig. 3.1. Semi-schematic view of a simple spin-polarization detector based on low energy electron diffraction from a single crystal surface (W(001)). Note the scale

plane, the intensities of the (2,0) and the ($\bar{2}$,0) beams are no more the same. Their normalized intensity difference is, as explained in more detail below, a measure of the magnitude of the polarization of the incident beam. The kinetic energy of the electrons at the crystal preferably remains fixed (otherwise the detectors would have to be moved around). As the energy of the electrons to be analyzed needs not to be the same as the scattering energy, an electrostatic lens system is provided that focuses the beam and brings the electrons to the scattering energy (104.5 eV was chosen in this case). This system is well suited for elastic scattering investigations as no electrons with energy higher than the primary energy are present in the beam at the detector. For inelastic scattering or for emission spectroscopies an energy dispersive analyzer has to be inserted in front of the spin detector. The LEED detector allows for single electron counting, in contrast to the absorbed current detector described below. With suitable channeltrons count rates up to 10^6 counts per second can be measured reliably. The detector is small (note the scale in Fig. 3.1) and operates at low voltages.

Calibration

Every polarization detector needs to be calibrated. One way of doing this is to use a primary beam of known polarization. In order to keep the uncertainty small, the polarization should be as high as possible. Photoionization of alkalis provides a low intensity beam nearly 100 % polarized (see e.g. Kessler /1976/), but this source represents a rather elaborate experiment in itself. A more convenient way is to make a double diffraction experiment using two equal crystals and an unpolarized electron source /Kirschner and Feder, 1979/. This experiment offers a self-calibration capability, independent of theoretical data (except for the sign of the asymmetry).

In general terms, the polarization $P_{\underline{k}_0\underline{k}_1}$ of a diffracted beam with wavevector $\underline{k}_1$, using an unpolarized primary beam of wavevector $\underline{k}_0$ is given by (2.18):

$$P_{\underline{k}_0\underline{k}_1} = \frac{\text{tr}(\underline{S}^{\dagger}_{\underline{k}_0\underline{k}_1}\underline{S}^{\dagger}_{\underline{k}_0\underline{k}_1}\sigma)}{\text{tr}(\underline{S}_{\underline{k}_0\underline{k}_1}\underline{S}^{\dagger}_{\underline{k}_0\underline{k}_1})} = \frac{\text{tr}(\rho_{\underline{k}_0\underline{k}_1}\sigma)}{\text{tr}\,\rho_{\underline{k}_0\underline{k}_1}} . \tag{3.1}$$

If this beam is used as the primary beam in a second diffraction experiment, the intensity of the diffracted beam $\underline{k}_2$ is proportional to:

$$I_{\underline{k}_2} \sim \frac{\text{tr}(\rho_{\underline{k}_1\underline{k}_2}(P_{\underline{k}_1\underline{k}_2}))}{\text{tr}\,\rho_{\underline{k}_0\underline{k}_1}} = \frac{\text{tr}(\underline{S}_{\underline{k}_1\underline{k}_2}(1+P_{\underline{k}_0\underline{k}_1}\sigma)\underline{S}^{\dagger}_{\underline{k}_1\underline{k}_2})}{\text{tr}(1+P_{\underline{k}_0\underline{k}_1}\sigma)} . \tag{3.2}$$

The polarization $P_{\underline{k}_1\underline{k}_2}$ of the beam $\underline{k}_2$ is given by (2.17):

$$P_{\underline{k}_1\underline{k}_2} = \frac{tr(\underline{S}_{\underline{k}_1\underline{k}_2}(1+\underline{P}_{\underline{k}_0\underline{k}_1}\sigma)\ \underline{S}^{\dagger}_{\underline{k}_1\underline{k}_2}\sigma)}{tr(\underline{S}_{\underline{k}_1\underline{k}_2}(1+\underline{P}_{\underline{k}_0\underline{k}_1}\sigma)\ \underline{S}^{\dagger}_{\underline{k}_1\underline{k}_2})} \quad . \tag{3.3}$$

It is evident from (3.2) that the intensity of the diffracted beam $\underline{k}_2$ depends on the polarization of the incident beam $\underline{k}_1$, which provides the basis for the self-calibration procedure.

The above general relations may be strongly simplified when applied to a special case used in practice by exploiting the symmetry arguments of Sect. 2.4.2. We assume the scattering plane for the first and second diffraction to be identical and to coincide with the same mirror planes of the two crystals. In this case the polarization vector $\underline{P}$ and the asymmetry vector $\underline{A}$ are normal to the scattering plane and we have $\underline{P} = \underline{A}$ for each of the two diffraction processes. It is sufficient to consider only the magnitudes of the vectors in this case. For the self-calibration the scattering conditions for each diffraction have to be the same. As we consider the calibration of the detector of Fig. 3.1 we assume normal incidence of the primary beam in both cases. This geometry is depicted in Fig. 3.2. Beams other than the (2,0) beams are omitted. The source beam is unpolarized, which can be thought of being composed of two totally but oppositely polarized beams of unity intensity. The length of the arrows indicates the relative intensity of the electrons with opposite spin orientations. The intensity loss after diffraction is

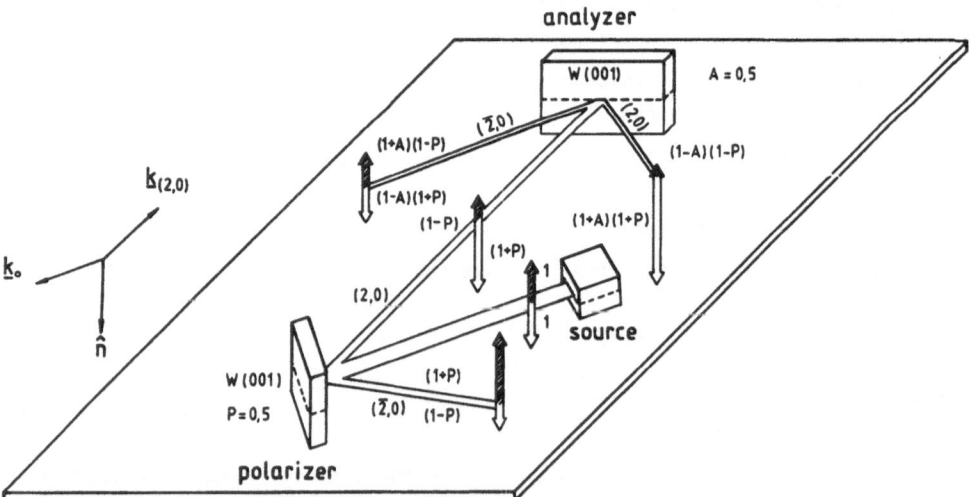

Fig. 3.2. Schematic view of a double diffraction experiment with self-calibration capability. The unpolarized electrons from a thermal source are polarized at the first W(001) crystal (polarizer), visualized by unequal lengths of the up and down arrows after scattering. The (2,0) beam is diffracted a second time at an identical crystal. The intensities of the two beams diffracted from the analyzer crystal are different, as indicated by the total length of the arrows, which provides a measure of the polarization of the beam incident onto the analyzer. If the same geometry and energy is used for polarizer and analyzer, if follows P = A and the intensity difference yields the square of P (self-calibration)

indicated by the widths of the beams. After the first diffraction the (2,0) and
($\bar{2}$,0) beams are each polarized with polarization P. From the diagonal form of the
density matrix $\rho_{k_{-0}k_1}$ (2.13) we obtain for the relative numbers of electrons with
spin-up ($N_\uparrow^1$) and spin-down ($N_\downarrow^1$):

$$\frac{N_\uparrow^1}{N_\downarrow^1} = \frac{1 \cdot (1 + P)}{1 \cdot (1 - P)} = \frac{1 + P}{1 - P} \tag{3.4}$$

where the unity stands for the relative intensity of the primary electrons with
the corresponding spin orientation. We briefly recall the definition of the sign
of $\underline{P}$: The polarization vector is positive if it points into the same direction as
the normal $\underline{n}$ to the scattering plane (2.35). For the (2,0) beam $\underline{n}$ points downwards
while for the ($\bar{2}$,0) it points upward. Therefore the two beams have the same polarization with respect to sign and magnitude, but the relative intensities of the
polarized electrons with respect to the scattering plane are just inverted.

After the first diffraction there is no intensity difference between ($\bar{2}$,0) and
(2,0) beams (indicated by the same total length of the arrows) because the primary
beam is unpolarized. The (2,0) beam from the first diffraction impinges normally
onto the analyzer crystal, characterized by the asymmetry A. As the asymmetry
equals the polarization in this geometry, we obtain from (3.2) or from (3.4) for
the ratio of the spin-up and spin-down intensities in the (2,0) beam:

$$\left.\frac{N_\uparrow^2}{N_\downarrow^2}\right|_{(2,0)} = \frac{N_\uparrow^1 (1 + A)}{N_\downarrow^1 (1 - A)} = \frac{(1 + P)(1 + A)}{(1 - P)(1 - A)} \tag{3.5a}$$

and in the ($\bar{2}$,0) beam:

$$\left.\frac{N_\uparrow^2}{N_\downarrow^2}\right|_{(\bar{2},0)} = \frac{N_\uparrow^1 (1 + A)}{N_\downarrow^1 (1 - A)} = \frac{(1 - P)(1 + A)}{(1 + P)(1 - A)} \quad . \tag{3.5b}$$

The polarization of the beams after the second diffraction is

$$P_{(2,0)} = \frac{(1 + P)(1 + A) - (1 - P)(1 - A)}{(1 + P)(1 + A) + (1 - P)(1 - A)} = \frac{A + P}{1 + AP} \tag{3.6a}$$

$$P_{(\bar{2},0)} = \frac{(1 + A)(1 - P) - (1 - A)(1 + P)}{(1 + A)(1 - P) + (1 - A)(1 + P)} = \frac{A - P}{1 - AP} \quad . \tag{3.6b}$$

Now the polarization of the (2,0) and the ($\bar{2}$,0) beams are no more equal. For the
case chosen in our example (P = 0.5; A = 0.5) the (2,0) beam is polarized by +80 %
while the ($\bar{2}$,0) beam is unpolarized.

Furthermore the intensities are no more equal, which opens the way to measure P
by means of a measurement of relative intensities of beams (2,0) and ($\bar{2}$,0) from
the analyzer crystal. The normalized difference D of the two beams is obtained
from (3.2) for the general case or from (3.5) for our special case:

$$D = \frac{I_{(2,0)} - I_{(\bar{2},0)}}{I_{(2,0)} + I_{(\bar{2},0)}} = \frac{(1+AP)-(1-AP)}{(1+AP)+(1-AP)} = AP \quad . \tag{3.7}$$

Thus P may be determined from the normalized intensity difference of the two detector beams, provided A is known. The same relation may be used for the self-calibration of the double diffraction experiment. If the scattering conditions are the same at the polarizer crystal as at the analyzer crystal, we have A = P and the normalized intensity difference D_{cal} at the calibration point equals the square of the polarization:

$$D_{cal} = P^2 \quad \text{or} \quad A = P = \pm\sqrt{D_{cal}} \quad . \tag{3.8}$$

This calibration procedure is valid for all energies and all beams, provided the scattering geometry is the same at both crystals. The sign of the polarization is lost, however, due to the squaring. It may be obtained from a polarized source, or by comparing to data measured with a Mott detector, or from a theoretical LEED calculation. As expected, all three alternatives yielded the same result (Kirschner and Feder /1979/, Wendelken and Kirschner /1981/, O'Neill et al. /1975/, Wang et al. /1981/). In the actual calibration use was made of the time reversal symmetry for the (2,0) beam from W(001), which allows to interchange the normally incident primary beam with the outgoing (2,0) beam. For this purpose the polarizer crystal is set parallel to the analyzer crystal, which can be done with high accuracy, and the electron gun is swept around the polarizer crystal at fixed energy until the (2,0) beam leaves the crystal along the surface normal. As only one mechanical rotation is involved in this procedure it leads to improved accuracy.

Once the calibration has been carried out with an arbitrary set of parameters, the analyzer part may be regarded as a 'black box' and a search for better parameters is conducted at the polarizer crystal. In this way the intensity and polarization properties of the 'detector beams' have been studied as a function of energy and angle of incidence. From a good polarization detector one expects high polarization sensitivity, expressed by the asymmetry A, and high efficiency expressed by the ratio of the detected intensity I to the incident current I_0. Assuming the background to be negligible, the error ΔP in the measurement of the polarization P of the incident beam is determined by the counting statistics and the asymmetry A. Assuming further Poisson statistics to be applicable, the 1σ error of the polarization is /Kessler, 1976/

$$\Delta P = [1/(NA^2)]^{1/2} \tag{3.9}$$

where N is the total number of counts $N = (N_{(2,0)} + N_{(\bar{2},0)})$. For a given accumulation time N is determined by the efficiency I/I_0, which means that the expression

$$F = A^2 \cdot \frac{I}{I_0} \tag{3.10}$$

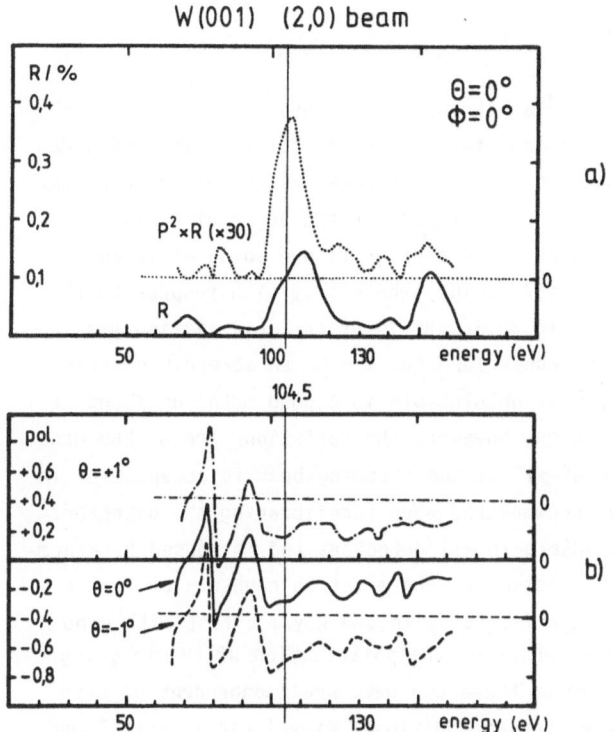

W(001) (2,0) beam

a)

b)

Fig. 3.3. a) Experimental data for the reflectivity R and the figure of merit P^2R;
b) The polarization of the detector beam and its sensitivity with respect to the
angle of incidence Θ. The "working point" at 104.5 eV is indicated

should be maximum. This "figure of merit" has originally been developed for Mott
detectors but is also applicable to the LEED detector. Though the asymmetry enters
quadratically, a high number of A is useful only at sufficient intensity. The relevant data for the (2,0) 'detector beam' are shown in Fig. 3.3. The measured reflectivity $R = I_{(2,0)}/I_0$ and the figure of merit P^2R at normal incidence are shown
in Fig. 3.3a while the polarization curves for three angles of incidence Θ in the
[1,0] azimuth are given in Fig. 3.3b. The reflectivity has a maximum near 110 eV,
which is close to a Bragg peak, with a shoulder at 105 eV. The reflectivity is
0.11 % at this energy. The polarization curve for Θ = 0° exhibits a strong plus-
minus feature near 80 eV and a relatively flat behaviour around 105 eV. The optimum working point of the detector may be read from the curve P^2R in Fig. 3.3a. It
is evident, that the figure of merit is not maximum for the largest polarization
values, but rather around 105 eV, in spite of a substantially smaller polarization. The second intensity maximum near 155 eV, though of similar intensity is not
useful because of lack of polarization. From these results the scattering energy
at the analyzer crystal was fixed at 104.5 eV. The polarization sensitivity was
found to be A = 0.27 ± 0.02. The efficiency is

$$I/I_0 = \frac{I_{(2,0)} + I_{(\bar{2},0)}}{I_0} = 2.2 \cdot 10^{-3}$$

and the figure of merit is $F = 1.6 \cdot 10^{-4}$ at the working point.

The figure of merit certainly is an important criterion for a polarization detector, but by no means the only one. Of equal importance are the background, the apparatus asymmetry and the sensitivity with respect to the angle of incidence. The background in the present case normally is negligible due to an efficient shielding of the channeltrons (~ 0.1 cts/s). The sensitivity with respect to the angle of incidence is illustrated in Fig. 3.3b. At 79 eV the polarization goes from strongly positive for $\Theta = +1°$ to about zero for $\Theta = 0°$ to strongly negative for $\Theta = -1°$. Such a behaviour clearly is untolerable in a good detector /Wang et al., 1981/. For the working point chosen, however, the variations are of the order of a few percent. Thus a divergence of $\pm 2°$ of the incoming beam is acceptable. In principle larger divergence may be accommodated when recalibrating the detector.

The apperatus asymmetry is unavoidable in all detectors and is caused by a number of sources e.g. unequal detection losses of the electron counters or slight misalignment. It may be determined experimentally in two ways: a test with unpolarized primary electrons and/or a reversal of the polarization of the incoming beam. The apparatus asymmetries found in these two ways are independent of each other and have to agree, which allows for an additional experimental control and an improvement of the accuracy.

A particular advantage of the LEED detector is its small size and weight. These properties allow to use it as an 'add-on' to existing electron spectrometers and to develop an angle-, energy- and spin-resolving electron spectrometer system. This may be applied to a variety of spectroscopies, such as photoemission, Auger spectroscopy or secondary electron spectroscopy. One of the more advanced momentum- and spin-resolving systems to date, capable of measuring all three components of the polarization vector is shown schematically in Fig. 3.4 (Oepen et al. /1983/, Oepen /1984/). The electrons from the sample are accepted by a zoom-lens, accelerated or decelerated to the chosen pass energy and focused into the spectrometer. The electrostatic spectrometer of the CMA-type with nearly second order focusing operates with a virtual entrance slit, which allows to pass the exciting radiation through the whole system onto the sample. After energy analysis the electrons are fed into the spin polarization detector where two components of the polarization vector are measured simultaneously. Instead of using a second pair of channeltrons for the (0,2) and (0,$\bar{2}$) beams a channelplate is placed behind the retarding grids that accepts a large number of diffracted beams. The collector is of the resistive anode type which provides an electronic x-y readout of single electron diffraction events (Stair /1980/). Electronic windows are set around the diffraction spots and the beam intensities are measured in a quasi-parallel way. In the geometry shown the (2,0)/($\bar{2}$,0) beam pair detects the transverse polarization

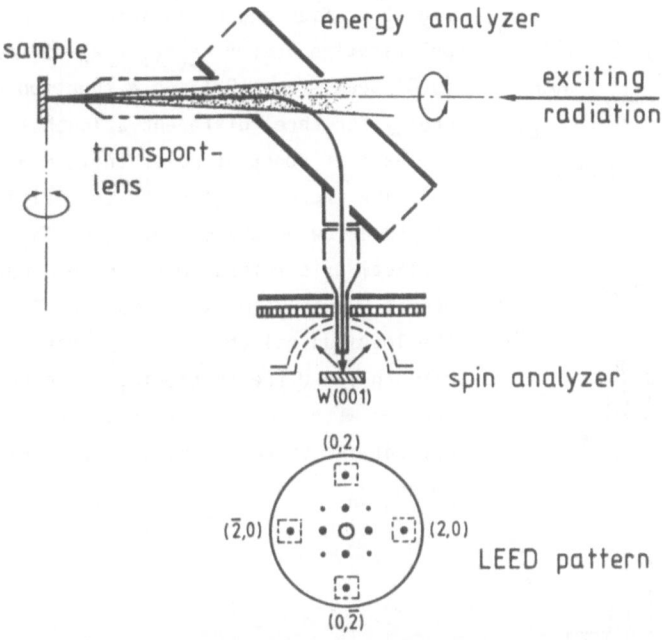

energy analyzer

sample

exciting
radiation

transport-
lens

spin analyzer

W(001)

(0,2)

(2̄,0)   (2,0)

LEED pattern

(0,2̄)

Fig. 3.4. A momentum- and spin-resolving electron spectrometer system, as for example, used in photoemission with circularly polarized synchrotron radiation. The spin analyzer is attached to the electrostatic energy analyzer and is moved with it about two perpendicular axes. The LEED pattern is observed on an oscilloscope with single electron counting

component normal to the optical axis of the zoom lens and normal to the paper plane. The $(0,2)/(0,\bar{2})$ beam pair detects the longitudinal polarization component along the optical axis. Of the remaining beams, the $(1,1)$ beams measure independently the projections at 45°. The $(1,0)$ beams, which have small polarization sensitivity at the working point serve to control the apparatus asymmetry. The complete spectrometer system is rotatable about a vertical axis lying in the sample surface. The second transverse polarization component, lying in the paper plane, can also be measured when the spectrometer (including the polarization detector) is rotated out of the paper plane about the optical axis. The longitudinal component is measured redundantly in this position. This spectrometer system has been used in spin-polarized photoemission from non-magnetic solids using circular polarized synchrotron radiation (see Chap. 4). A test of its performance has been made for elastic scattering of unpolarized electrons from W(001). The electron gun was in the paper plane and the $(0,0)$ beam was measured at a polar angle of $\Theta = 28°$ for various azimuthal angles ϕ between the scattering plane and a mirror plane of the crystal. The orientation and magnitude of the polarization vector projected onto a plane normal to the scattering plane are shown in Fig. 3.5 as a function of energy. At $\phi = 0$ (Fig. 3.5c), i.e. the mirror plane in the scattering plane, the polarization vector is normal to the scattering plane and no longitudinal component is detected. For azimuthal angles $\phi = -5°$ (Fig. 3.5d) and $\phi = +2.5°$ (Fig.

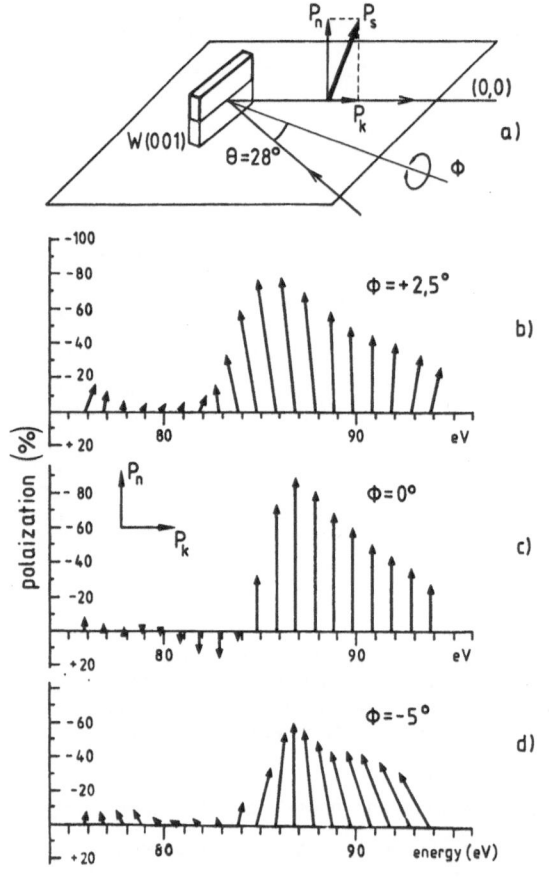

Fig. 3.5. Size and orientation of the polarization vector P_s (see a) of the (0,0) beam from W(001) as a function of energy for three different azimuthal angles ϕ at constant polar angle $\Theta = 28°$. The energy is given by the origin of the arrow on the energy axis. The vertical axis refers to the normal component P_n; the same scale is used for the longitudinal component P_k. For the azimuth $\phi = 0$ the scattering plane coincides with a mirror plane of the crystal and there exists only the component P_n

3.5b) the polarization vector contains positive or negative longitudinal components, antisymmetric with respect to $\phi = 0°$. The second transverse component, presumably small as the longitudinal component is not far from $P_\perp$ as defined in Fig. 2.7, was not measured in this test. This example nicely demonstrates the influence of multiple scattering effects on the polarization vector, as discussed in Sect. 2.4. We remember that in the kinematic limit the polarization vector would be normal to the scattering plane, irrespective of the azimuthal angle ϕ.

The LEED detector being a relatively new development, a few comparisons shall be made with the conventional Mott detector. The latter one usually operates around 100 kV, which makes it bulky (characteristic dimensions 2-3 m) and expensive. In some new developments (Hodge et al. /1979/, Gray et al, /1984/, Koike and Hayakawa /1983/) a size reduction to dimensions of a few ten centimeters was achieved through special insulation techniques, and/or reduction of the high voltage. The typical figure of merit of Mott detectors is around 10^{-5}, about an order of magnitude lower than the LEED detector. A figure of merit of $1 \cdot 10^{-4}$ has been reported /Heinzmann et al., 1975/ exhausting nearly all possible improvements, and marking the limits of this detector line. Even the present prototype LEED detector provides a slightly superior figure of merit, though only a modest effort went into its development. No systematic search for more efficient parameter combinations has been carried out so far, though there are numerous possibilities: using other crystal planes or other materials, adsorbate systems or compound crystals. Reflectivities of more than 1 % have frequently been observed in LEED and polarization values around 50 % are not unusual /Erbudak et al., 1982/. Even if one does not hope to find these conditions together (they would improve the figure of merit by two orders of magnitude), a conservative estimate would consider an improvement by one order of magnitude realistic.

A certain restriction for the application of the LEED detector is its dependence on UHV conditions. A clean surface is contaminated by absorption of residual gases, with the rate depending on the pressure and the sticking coefficient. Under high vacuum conditions the contamination rate is too high for all materials, even for very low sticking coefficients. If a clean crystal surface is used for the detector, the noble metals like Au or Pt are preferable in principle, but the sticking coefficient is not the only criterion. Since 'inert' surfaces also become contaminated after some time, it is equally important to have simple cleaning procedures available that work fast and without further control of the surface cleanliness. From this point of view W is preferable, in spite of its high sticking coefficient, as a flash to ~ 2500 K removes all adsorbed gases and restores the crystal to its previous state. During routine operation in the 10^{-10} Torr range a flash is made every 15 to 30 min, which interrupts the measurement for about 1 min.

3.1.2 The Adsorbed Current Detector

It is well known that the net current to a sample bombarded by energetic electrons may be zero. This is the case if the incident electron flux equals the flux of electrons leaving the sample. Besides elastically and inelastically scattered electrons this flux mainly consists of secondary electrons. In general there are two such zero crossings for conductors as a function of the primary energy. The first one at energies of several 100 eV occurs because with increasing energy there are more and more secondaries generated per primary electron. The second one appears at several keV to some 10 keV because the range of the primaries increases with increasing energy, the secondaries are produced at larger depth, and fewer electrons arrive at the surface with sufficient energy to overcome the surface barrier. The precise location of these zero crossings depends on a number of parameters, e.g. the material, the angle of incidence, the surface roughness, and the work function. It was first observed by Siegmann et al. /1981/ for a ferromagnetic glass that the low energy zero crossing also depends on the orientation of the electron spin of the primaries relative to the magnetization. This effect was attributed to the spin dependence of the elastic scattering from ferromagnets via exchange interaction. It cannot be ruled out, however, that the spin dependence of the inelastic mean free path plays a role. It may be small, it is true, but the effects observed are of the same order of magnitude. The same phenomenon was shortly afterwards observed on non-magnetic single crystals of Au /Erbudak and Müller, 1981/ and W /Celotta et al., 1981/. In this case it is the spin-orbit interaction which causes the effect. Subsequently the effect was observed also for polycrystalline materials (Erbudak and Ravano /1981/, Pierce et al. /1981/). Though it is very likely that inelastic scattering may contribute to this effect /Ravano et al., 1982/, in addition to elastic scattering, the origin of the phenomenon has not yet been understood in detail. Nevertheless, the effect has been utilized for another type of spin polarization detectors. The spin-dependent absorption effect has very clearly been demonstrated on Au(110) by Erbudak and Müller /1981/. Figure 3.6 shows a typical result for the plane of incidence normal to the [1$\bar{1}$0] direction. The upper part of the figure shows the energy value E_o at which the zero crossing of the sample current occurs for unpolarized primary electrons as a function of the angle of incidence Θ. The insert shows at $\Theta = 56.5°$ the absorbed current as a function of energy for up-spin ($I_\uparrow$), down-spin ($I_\downarrow$), and unpolarized primary electrons (dashed). Evidently the zero-crossing for unpolarized electrons ($E_0 = 118.4$ eV) splits up into two crossings displaced by about ± 1 eV with respect to E_0. The corresponding curves are not necessarily straight lines and they are not necessarily displaced parallel to each other, though they may appear as such over a small energy interval. The lower part of Fig. 3.6 shows the current difference $I = I_\uparrow - I_\downarrow$ relative to the primary current I_0 at the energy E_0

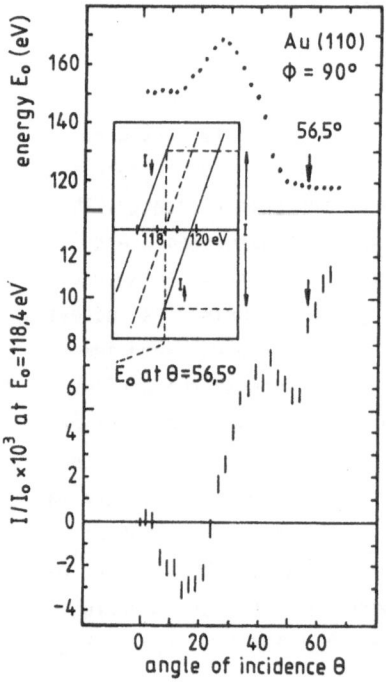

Fig. 3.6. Spin-dependent electron absorption on Au(110). The upper panel shows the energy E_0 as a function of the angle of incidence Θ at which the absorbed current is zero for unpolarized primary electrons. The insert shows the splitting of the zero crossing (at fixed Θ) for oppositely polarized incident electrons. The magnitude of the current $I = I_\uparrow - I_\downarrow$ relative to the incident current I_0 is shown in the lower panel as a function of the angle of incidence at $E_0 = 118.4$ eV

as a function of the angle of incidence. For the parameters chosen I/I_0 is of the order of 10^{-3} to 10^{-2}. After a calibration either by a primary beam of known polarization or a comparison to another polarization detector this intensity difference can be used as a measure for the polarization of the incoming beam. The statistical uncertainty of a polarization measurement has to be evaluated in a different way than for the LEED detector. The fluctuations of the current I are not determined by the current I itself but rather by the fluctuations of the incident beam and by the statistics of the secondary electron generation. This is evident from a measurement with unpolarized electrons at E_0. The average net current to the sample is zero, but the fluctuations of this current are far from zero. As shown by Pierce et al. /1981/ the figure of merit of this detector is

$$F' = \frac{1}{2\delta^2} \left(\frac{I}{I_0}\right)^2$$

where δ^2 is the variance of the statistics of secondary electron generation. Assuming Poisson statistics to be valid, the variance equals the mean value and we have $\delta^2 \approx 1$ as the secondary electron yield is close to unity. The largest I/I_0 reported so far in actual application of the detector is $I/I = 1.5 \cdot 10^{-2}$ /Erbudak and Ravano, 1983/ for a polycrystalline Au surface. Then the figure of merit is $F' = 1.1 \cdot 10^{-4}$, which is less than but comparable to that of the LEED detector described in the previous section.

Absorbed current detectors exploiting the spin-orbit interaction are sensitive to the polarization component normal to the plane of incidence. Therefore the two transversal components of the polarization vector can be obtained by rotating the detector assembly about the incident beam /Erbudak et al., 1982/. These components can, however, only be measured consecutively, not simultaneously as by the LEED detector.

The working point E_0 and the energy splitting of the two zero crossings was observed to be subject to drifts, most likely caused by contamination from the residual gas. Therefore re-calibrations are necessary, at least occasionally. The detector is sensitive to the angle of incidence, the energy, and the energy spread of the incident beam. Each of these factors must be handled the same way at the time of calibration and at the time of measurement /Pierce et al. 1981/. An advantage of this detector is that it is not bound to single crystal surfaces. A contaminated surface may be renewed by evaporating a fresh layer of e.g. gold, followed by annealing and recalibration. On the other hand relatively large primary currents are required that have to be two to three orders of magnitude above the useful range of operation of sensitive electrometers. A major drawback of the absorbed current detector is that single electron counting is not possible, a feature that is most desirable at low intensity levels. Therefore it appears most useful in applications where a high intensity source of known polarization is available.

3.2 Sources of Polarized Electrons

A number of polarized electron sources have been described in the literature. They
may be roughly classified into atomic physics sources and solid state sources,
which to some extent also describes their main area of application. In atomic
physics sources have been built on the basis of spin-orbit interaction in electron-atom scattering /Jost et al., 1981/, or by ionization of polarized or unpolarized atoms. For example the photoionization of unpolarized alkali atoms by circular light was used /Wainwright et al., 1978/ or the photoionization of polarized
alkali atoms by unpolarized light /Alguard et al., 1979/. As an alternative to
photoionization the chemiionization has been exploited: metastable He atoms, generated by optical pumping emit polarized electrons when interacting with N_2 or CO_2
/Keliher et al., 1975/. High polarization has been reported /Alguard et al., 1979/
as well as high current up to microamperes /Keliher et al., 1975/ but not both together. In principle, atomic physics sources are applicable to surface studies
too, but in general there is an atomic beam involved in one form or another, which
causes difficulties to establish UHV conditions. For reviews see Kessler /1976/,
Celotta and Pierce /1980/. Among the solid state sources the GaAs source is the
most widely used and will be discussed in some detail.

3.2.1 The GaAs Source

The GaAs source was proposed by Garwin et al. /1974/ and Lampel and Weisbuch /1975/,
and investigated experimentally by Pierce and Meier /1976/. Its operating principle
is the photoemission with circular polarized light, i.e. the 'operator effects' in
the language of Sect. 2.5. GaAs is a direct band gap semiconductor and the symmetry
adapted basis functions at the Γ point are predominantly of p character in the valence band and of s character in the conduction band. The relevant section of the
band structure is shown in Fig. 3.7. The p level is split by $\Delta = 0.34$ eV due to
spin-orbit coupling. The upper $p_{3/2}$ level is fourfold degenerate at Γ, while the
lower $p_{1/2}$ level is twofold degenerate. The schematic on the right hand side in Fig.
3.7. represents the energetic levels at Γ with the quantum numbers j according to
the projections of the total angular momentum on the quantization axis. When photons
with energy equal to the gap energy $E_g = 1.52$ eV (at 77 K) are absorbed there are
direct transitions possible from $p_{3/2}$ to $s_{1/2}$ ($\Delta \ell = -1$). For right circular light
σ^+ there are only transitions $m_j = -3/2 \rightarrow m_j = -1/2 \rightarrow m_j = +1/2$. The electrons in
$m_j = -1/2$ are oppositely polarized, so that a net polarization is observed only if
the transition probabilities are different. Explicit evaluation of the matrix elements, using only the angular parts, yields the relative intensities 3 and 1, respectively.
Therefore the polarization is -50 %. For isolated transitions from the split-off
band $p_{1/2}$ for $h\nu = 1.86$ eV the polarization would even be +100 %, but these are
masked by simultaneous transitions off the Γ point and the polarization is strong-

74

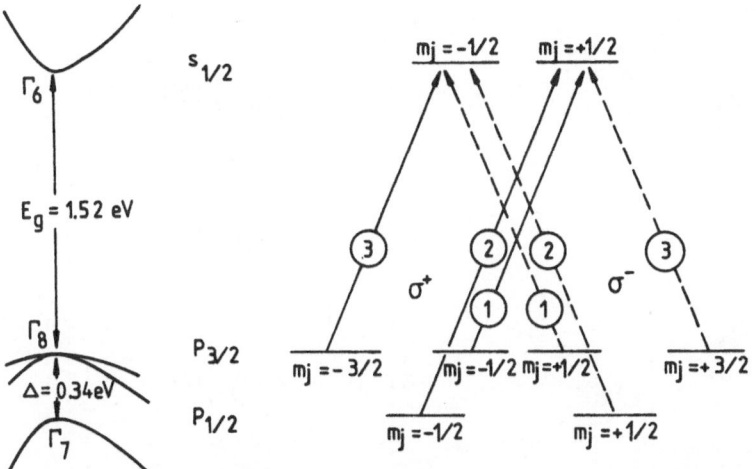

Fig. 3.7. Section of the bandstructure of GaAs around the Γ point (a) and transitions induced by right-circular (σ^+) and left-circular light (σ^-) across the band
gap (b). The calculated transition rates lead to a polarization of $\pm$ 50 % for
light with $h\nu$ = 1.52 eV

ly reduced. This 'internal photoeffect' generates polarized electrons in the conduction band, which can be observed indirectly through the circularly polarized
recombination radiation /Fishman and Lampel, 1977/. They can, however, not be observed in the vacuum for clean GaAs surfaces because the vacuum level lies above
the lower edge of the conduction band ('positive electron affinity'). To obtain
polarized electrons in the vacuum, a trick may be used that has been developed for
the production of extremely sensitive photocathodes: the crystal is strongly p
doped and covered by a thin layer of Cs or, more efficiently, by a layer of Cs +
Oxygen. Together with the band bending the low work function of the layer pulls
the vacuum level below the conduction band edge and 'negative electron affinity'
is achieved. The physics of negative electron affinity devices /Bell, 1973/ is under current investigation as the precise nature of the overlayer is not very well
known (Spicer et al. /1979/, Spicer et al. /1980/, Su et al. /1983/), and there
still is a bit of 'black magic' involved in the preparation of high quality cath-
odes /Pierce et al., 1980/.

From the experimental point of view GaAs is not ideal because of the small band
gap requiring infrared light. There are suitable light sources in the form of
GaAlAs laser diodes, but they do not produce a well collimated beam. Therefore
relatively large infra-red optics are necessary. Also, it is not trivial to focus
the invisible light to the desired place on the cathode surface and to change the
optical polarization without intensity variations. On the other hand, the HeNe
laser is a convenient, inexpensive and well collimated light source. The photon
energy, however, is 1.96 eV, which induces transitions from the split-off band in
GaAs and lowers the polarization drastically. An elegant solution is to increase

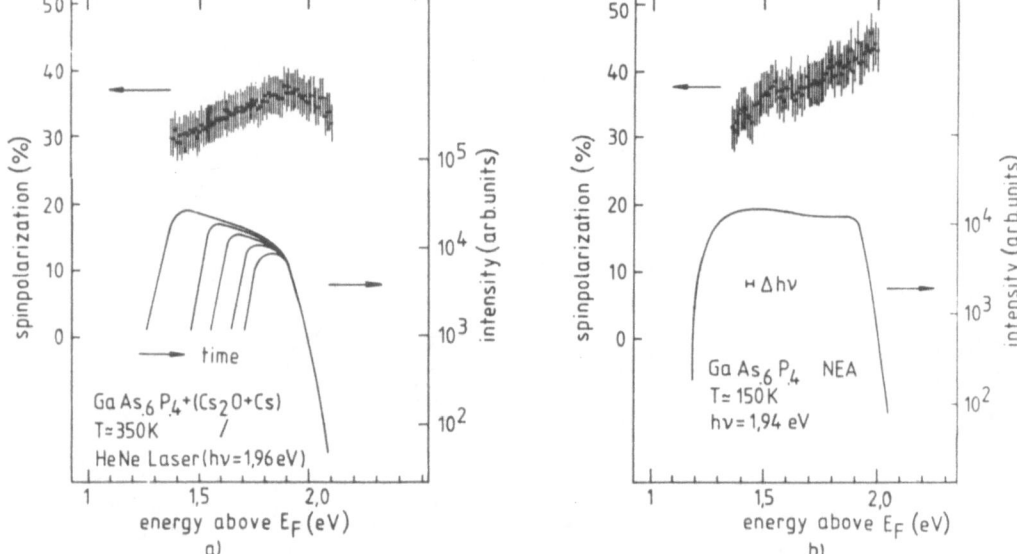

Fig. 3.8. a) Intensity distributions on a logarithmic scale as a function of time after activation (at T ≈ 350 K; HeNe laser). The shoulder at the right-hand side corresponds to emission from the conduction band, the left-hand edge represents the low energy cut-off due to the work function. The spin polarization distribution (linear scale) corresponds to the freshly activated cathode. The intensity distributions have been taken at different time intervals. The narrowest distribution corresponds to about 1 h after activation. To speed up degradation for these measurements the cathode was exposed to poor vacuum of ~ $1 \cdot 10^{-8}$ Torr.
b) Intensity (log scale) and spinpolarization (linear scale) distributions at T ≈ 150 K and hv = 1.94 eV after activation

the band gap by adding aluminum or phosphorous /Conrath et al., 1979/. Up to a P concentration of x = 0.45 the $GaAs_{1-x}P_x$ compounds remain direct band gap semiconductors /Reichert and Zähringer, 1983/. At x = 0.4 the band gap is in the range of 1.8 to 2.0 eV, depending on temperature, which matches the photon energy of the HeNe laser. This combination represents a very convenient source with good intensity. With a 15 mW laser emission currents of more than 100 µA have routinely been achieved in the author's laboratory.

Important properties of a polarized electron source are its energy distribution and the polarization over this distribution. For GaAs an energy width (FWHM) of 130 meV was reported /Pierce et al., 1980/ and a polarization of 42 % integrated over the energy distribution. For $GaAs_{0.6}P_{0.4}$ an integrated polarization of around 40 % was reported /Conrath et al., 1979/. A differential spin- and energy-analysis for GaAsP was made by Kirschner et al. /1983/ using a high resolution electron spectrometer and a LEED polarization detector as shown in Fig. 3.1. The main results are shown in Fig. 3.8a for a temperature of ~ 350 K and in Fig. 3.8b for ~ 150 K. The intensity distributions exhibit a trapezoidal or rectangular shape on the logarithmic scale. They are rather broad, up to 0.8 eV FWHM have been observed

at optimum activation, i.e. maximum negative electron affinity. The low energy
edge is observed to move with time, indicating an increasing workfunction due to
contamination of the activation layer. The high energy edge corresponds approximately to the conduction band edge and stays fixed. The polarization distribution
shows a maximum of ~ 38 % at 350 K, decreasing to both sides. At 150 K the polarization is maximum (~ 45 %) at the conduction band edge and decays towards lower
energies. The large width of the intensity distribution is attributed to energy
losses from optical phonon excitations in the bent-band region while the electrons
travel back and forth in the potential well until they finally escape into vacuum.
The decay of the polarization towards the low energy side was used to estimate the
spin-relaxation time (~ $2 \cdot 10^{-11}$s) and to discuss some of the competing depolarization mechanisms (Bir et al. /1976/, Fishman and Lampel /1977/). The difference in
the polarization maxima at 350 K and 150 K is explained by the temperature dependence of the band gap. While at 150 K the photon energy is well matched to the
band gap, the gap is smaller at 350 K. Therefore transitions off the Γ point contribute and the polarization decays towards higher energy. The depolarization was
found to be independent of temperature for epitaxially grown GaAsP crystals. Recently, a temperature dependent depolarization effect was observed with cleaved
GaAs crystals /Allenspach et al., 1984/. The larger energy width for GaAsP than
for GaAs is a direct consequence of the larger negative electron affinity. After
aging the cathode a width of less than 50 meV was observed, which was mainly determined by the spectrometer resolution. It should be noted that the degradation
time in Fig. 3.8a is not representative for operation in good vacuum of ~ 10^{-10}
mbar or less. Decay time constants of 10 h /Pierce at al., 1980/ or even 100 h
/Stocker, 1975/ have been obtained. The observation that the high energy cut-off
remains fixed on the energy scale, while the low energy edge depends only on the
work function, provides a means to generate a monochromatic beam of electrons at
fairly high current. As demonstrated by Feigerle et al. /1984/ a beam current of 1
µA at 30 meV energy spread can be produced by carefully adjusting the work function and matching the photon energy precisely to the band gap. This monochromatic
current is about an order of magnitude higher than that obtainable with a thermal
electron source and a monochromator. As the space charge limits are relaxed when
starting with a narrow energy distribution /Ibach and Mills, 1982/ there is hope
to improve further on the ratio of beam current to resolution by using a photocathode as the electron source in a monochromator system. Such a system would be
very useful also in conventional electron spectroscopy, with the spin polarization
as an additional bonus.

    While the intensity of this source may be satisfactory for most applications,
one would often like to have higher polarization. Though the theoretical limit of
50 % was reached /Alvarado et al., 1981a/ and even slightly exceeded under special
conditions /Pierce and Meier, 1976/ no break-through towards 100 % polarization
has yet been achieved. The recipe is simple, in principle: it should be sufficient

to lift the degeneracy of the $p_{3/2}$ level and to match photon energy and energy
width well enough. In chalcopyrite compounds the degeneracy is removed naturally
/Zürcher and Meier, 1979/, in GaAs it may be removed by uniaxial compression
/D'yakonov and Prerel', 1974/. Superlattices of GaAs/GaAlAs would have the same
effect /Alvarado et al., 1981b/ but the success of much experimental effort in
this area is still lacking. Possibly, molecular beam epitaxy with its ability to
'taylor' band-structure properties or to produce strained superlattices might
bring a solution /Biefeld et al., 1983/.

In the next section we will discuss some sources that yield higher polarization
(or promise to do so) at the expense of other disadvantages. One major advantage
of the GaAs source is the easy reversibility of the polarization vector. It is
sufficient to reverse the handedness of the light by standard optical means, without any change of the electron optics. This greatly facilitates the suppression of
apparatus asymmetries to the order of 10^{-3} or less.

3.2.2 Other Sources

A polarized electron source which combines high polarization with extreme electron
optical brightness is the field emission source. It was first shown by Müller et
al. /1972/ that electrons tunneling from a W tip through a layer of ferromagnetic
EuS at about 10 K into vacuum may be polarized up to 90 %. This effect was later
verified by Kisker et al. /1978/ and investigated in more detail. Though there are
still some open points, the process can essentially be understood in the following
way: The external electric field produces a potential gradient in the EuS layer
which pulls the conduction band below the Fermi level of W at some distance from
the W/EuS interface. The conduction band being exchange split, this crossing is
closer to W for the majority electrons than for minority electrons. As the tunneling probability depends in an extremely sensitive way on the wavefunction overlap
of W and EuS electron levels, the tunneling probability for electrons from W with
the majority spin orientation is greatly enhanced over that for the minority electrons. In this way the EuS layers act as an almost perfect 'spin-filter' for electrons tunneling from the Fermi level of the tungsten tip. The beam currents are
reported to be 10^{-8} A /Kisker et al., 1978/, with an extremely small electron optical source diameter (a few nm). With this source a fine, highly polarized electron beam can be formed that could be used in a polarized electron microscope.
However, there are some technological problems: producing the EuS coated tips,
controlled annealing, emission in a high magnetic field and cooling with liquid
helium. A major drawback of this source is that the polarization reversal has to
be accomplished by a reversal of the magnetic field. As the emitted electrons have
to pass through the magnetic field the electron optical properties are changed and
the beam spot's location and intensity may change. For these reasons the polarized
field emission source will probably be used only if the electron optical brightness is of primary concern.

There is also the possibility to polarize an electron beam from a thermal cathode by reflection from a high-Z single crystal surface. Polarizations up to 90 % may be achieved (see Fig. 3.5) at the expense of intensity. With an unpolarized beam current of 1 mA the polarized beam current could be of the order of 10^{-8} A. Higher currents of 0.1 µA could perhaps be obtained at a polarization level of 50 % /Erbudak et al., 1982/. The major drawback of this technique is the space charge saturation of the primary, unpolarized beam, not that of the polarized beam. On the other hand, such a source could be easily miniaturized (the space charge problems even leave no other choice), and the whole assembly could be used in much the same way as a conventional electron gun.

Another possibility should also be mentioned: exploiting initial state effects in photoemission from ferromagnetic materials. It has been observed by Kisker et al. /1980/ that near the photothreshold of Ni(110) for linear polarized light a spin polarization around -90 % exists. With conventional light sources the yield is too low to make a polarized electron source out of this effect. However, with frequency-doubled lasers and/or a slight lowering of the work function this effect may result in an alternative source with high polarization. The reversal of the polarization in this case also would have to be made by magnetization reversal, but the source would operate under zero external field conditions, and stray fields could probably be made negligible. This alternative remains as yet untested.

4. Results from Non-Magnetic Crystals

In this and the following chapter the results of electron scattering and electron emission experiments with magnetic and non-magnetic solids are presented. We shall treat spin-orbit and exchange interaction separately, though from a physical point of view this separation is somewhat artificial since both types of interaction occur in both types of materials. Interference between polarization effects due to spin-orbit interaction and those due to exchange interaction are discussed in Chap. 5.

4.1 Spin-Polarized Electron Diffraction

We start with an experimental test of the symmetry relations established in Sect. 2.4, namely the relation between the vectors of asymmetry $\underline{A}$ and polarization $\underline{P}$ and their dependence on crystal symmetries. A quantitative comparison between theory and experiment is made for W(001) and the sensitivity of the spin polarization to structural parameters is demonstrated. So-called 'surface resonances' in LEED are seen to be spin-dependent. Temperature effects are treated and some effects of adsorbates on the spin polarization are shown.

4.1.1 Symmetry Relations

When dealing with complex physical phenomena, it appears advisable to look for general symmetry properties that do not require heavy calculations. We discussed symmetry relations in Sect. 2.4 and found that spin-polarized LEED differs from conventional LEED by removing time reversal degeneracies. A very nice example of symmetry studies has been given by Bauer /1980/, Bauer et al. /1980/, Bauer et al. /1983/, Feder et al. /1980/ for the (00) beam from Pt(111). According to the results of ion scattering (Davies et al. /1978/, Van der Veen et al. /1979/) and LEED (Kesmodel et al. /1977/, Adams et al. /1979/) this surface is not reconstructed and the top layer distance is close to the bulk value. If the scattering plane coincides with a mirror plane of the crystal, we recall from Chap. 2 that the component $P_\perp$ of the polarization vector $\underline{P}$, which is normal to the crystal surface vanishes identically. Outside this configuration there may appear components of $\underline{P}$ in the scattering plane due to multiple scattering. In a rotation diagram

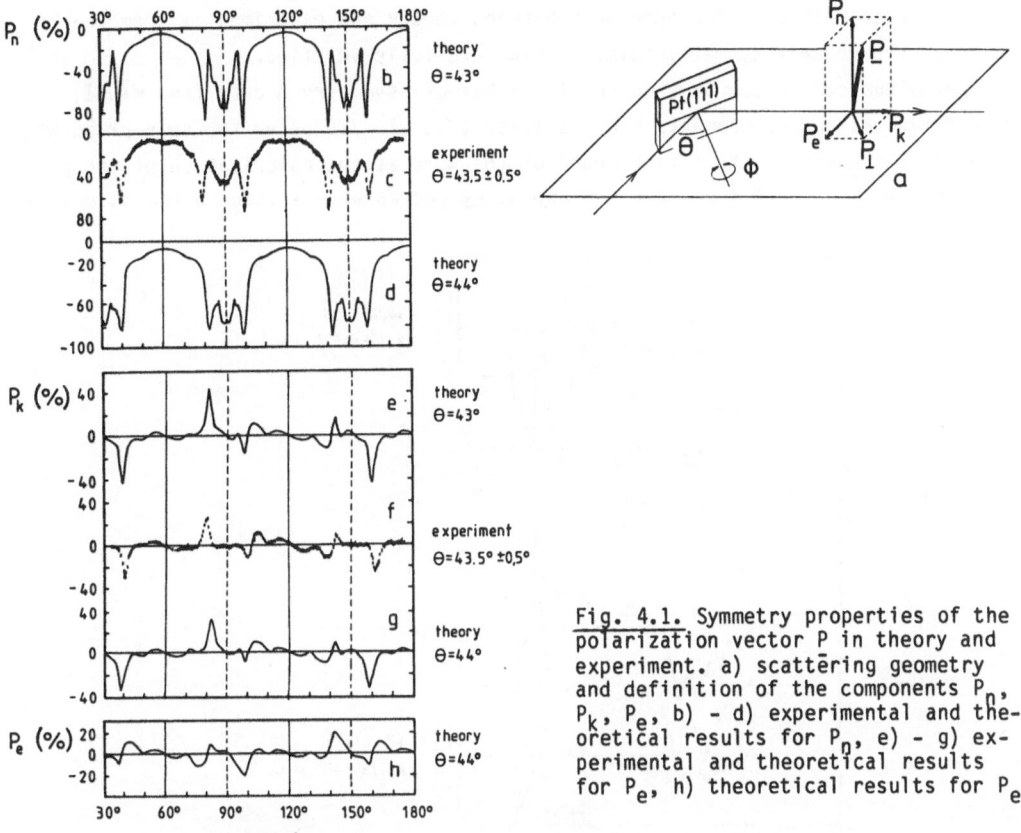

Fig. 4.1. Symmetry properties of the polarization vector P in theory and experiment. a) scattering geometry and definition of the components P_n, P_k, P_e, b) - d) experimental and theoretical results for P_n, e) - g) experimental and theoretical results for P_e, h) theoretical results for P_e

$\underline{P}(\phi)$ of the specular beam the polarization vector will therefore in general be inclined relative to the scattering plane. As the polarization vector has the transformation properties of an axial vector, the component $P_\perp$ will show an antisymmetric behaviour with respect to the mirror plane. In the experiment the polarization components normal to the scattering plane ($\underline{P}_n$) and along the axis of the outgoing beam ($\underline{P}_k$) were measured (see Fig. 4.1). The component P_n corresponds to $\underline{P}_\parallel$ in the language of Sect. 2.4, while P_k represents the projection of the component $P_\perp$ onto the scattered beam axis. The third component ($\underline{P}_e$) was not measured in this experiment involving a Mott detector. The points in Fig. 4.1 represent the experimental data for P_n and P_k, the lines represent the theoretical data. Looking at P_k and P_e we see that both components vanish in a mirror plane (at $\phi = 60°$ and $\phi = 120°$), i.e. $P_\perp = 0$. They behave antisymmetrically with respect to these symmetry lines, while the component P_n shows mirror symmetry, as expected. It is remarkable that the lines at $\phi = 90°$ and $\phi = 150°$ do not give rise to symmetric or antisymmetric structures, as would be the case for a surface of 6-fold symmetry. The intensity of the (0,0) beam was not measured in this experiment, but its 6-fold symmetry is well-known from conventional LEED studies (Woodruff and Holland /1970/,

Lagally et al. /1971/). The agreement between theory and experiment is remarkably good and the general symmetry relationships are fully veryfied.

A previous comparison by Wang et al. /1981/ of asymmetry $\underline{A}$ data from W(001) with polarization measurements $\underline{P}$ by Kalisvaart et al. /1978/ on the same plane was no proof for $\underline{A} = \underline{P}$ in the strict sense of the word as the polarization of the primary beam was not well known and the asymmetry curves were fitted to the polariza-

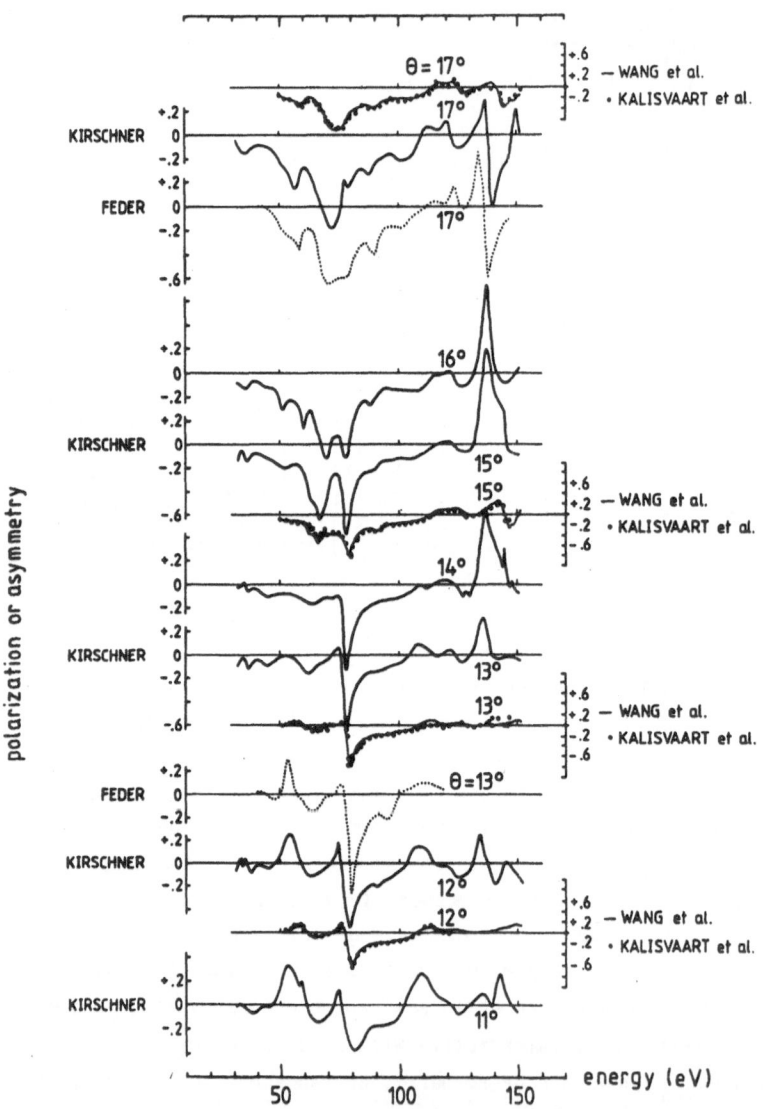

Fig. 4.2. A compilation of some experimental and theoretical polarization data for the (0,0) beam from W(001) as a function of energy. The data of Kalisvaart et al. /1978/ and Kirschner represent the polarization after scattering of an unpolarized beam. The data of Wang et al. /1981/ represent the intensity asymmetry after scattering for a completely polarized incident beam. The dotted lines (Feder, private communication) show theoretical results for which A = P

tion curves. The structures of the A(E) and P(E) curves do, however, agree quite well (see Fig. 4.2 below), confirming the general symmetry relations also in this case.

This apparently clear-cut state of affairs had been questioned by Reihl and Dunlap /1980/ and Reihl /1981/ on the basis of their observations on a Au(110) surface. They found that the (2x1) reconstructed surface in some cases did not show the agreement of $\underline{A}$ and $\underline{P}$, notably for the (0,1/2) beam. Within a kinematical model they concluded that this beam showed the reduced symmetry of the top layer, while the substrate beams show the 2-fold symmetry of the crystal. They excluded all reconstruction models having a mirror plane normal to the [110] direction or a two-fold symmetry. These conclusions were shown to be wrong by Müller et al. /1981/ by experimentally demonstrating the symmetries expected from the crystal symmetry and by verifying the equality of $\underline{A}$ and $\underline{P}$ for the (0,1/2) beam.

This example emphasizes the sensitivity of the spin polarization to structural parameters on the one hand, and its sensitivity to experimental errors on the other hand.

4.1.2 Structure Analysis

The detection and analysis of symmetries in physical systems certainly is esthetically satisfying. Symmetry properties can also be exploited experimentally; for example the self-calibration of the LEED detector relies on the time reversal symmetry and the symmetry properties of mirror planes. However, the results are of a qualitative nature only. For quantitative conclusions about surface structure properties the experimental results have to be compared to theoretical calculations. Because of strong multiple scattering in LEED it is not possible to extract parameters directly from measured data by a suitable transformation as is done in e.g. X-ray diffraction. There have been attempts to apply transform methods to averaged LEED intensity data (Adams and Landman /1927/, Marsh and King /1979a/), not all of which were successful (Adams /1979/, Marsh and King /1979b/). The spin polarization in LEED owes its sensitivity to structural parameters exclusively to multiple scattering. Any analysis based on the assumption of a kinematical model is therefore bound to fail. Even if some averaging procedure worked perfectly, i.e. if all multiple scattering effects were averaged out, one would not obtain any structural information from the polarization since intensity and polarization then are decoupled. The polarization would be that of the (muffin-tin) atom and the structure dependence would be lost /Wang et al., 1982a/.

The analysis procedure used to date is that of 'trial and error': Theoretical calculations based on a particular structure model and particular electronic parameters are compared to the experimental results. The structural and/or electronic parameters are varied until good agreement is obtained. The agreement is judged according to more or less subjective criteria. In conventional LEED so-called 'reliability factors' have been introduced, which express the agreement between two curves, according to some criteria, by a single number (Zanazzi and Jona /1977/, Marcus et al. /1975/). They have the great advantage to characterize large sets of data by a few numbers. They are not unique, however, and there is some arbitrariness involved in choosing the criteria by means of which two curves are compared to each other. In spin-polarized LEED there are in general two sets of data, polarization P and reflectivity R of which several combinations may be formed. Both can be considered separately, one may form $P \cdot R = R_\uparrow - R_\downarrow$, the reflectivity difference for the two spin orientations, or even use the 'figure of merit' $P \cdot R$. The latter one emphasizes the polarization too strongly and the sign is lost. The $P \cdot R$ criterion emphasizes the differential character of spin-polarized LEED and appears not unreasonable. However, the formation of this difference involves the multiplication of two sets of data which mostly are obtained independently, thereby adding the experimental uncertainties. For these reasons P and R are generally compared separately. The experience acquired so far indicates that the polarization data of several different groups are in slightly better agreement than the intensity data of the same groups (cf. Feder /1981/). A compilation is shown in Fig. 4.2. This agreement is rather astonishing in view of the sometimes dramatic sensitivity of the polarization to an angular variation by a few tenth of a degree and energy variations by a few tenth of an eV (cf. the discussion of Fig. 3.3). A possible explanation might be that the polarization is a normalized quantity into which the analyzer transmission does not enter, while the reflectivity is an absolute quantity, strongly subject to the source and detector characteristics (for a comparison of conventional LEED data see e.g. Read and Russell /1979/, Ignatiev et al. /1977/).

An extensive experimental and theoretical study of spin-polarized LEED on W(001) has been made by Feder and Kirschner /1981a/. This surface is relatively 'open', and a reversible reconstruction is observed for temperatures below 0°C (King and Thomas /1980/, Debe and King /1979/, Felter et al. /1977/, Lu and Wang /1981/, Melmed et al. /1979/). Above room temperature the LEED pattern shows no superstructure and there is only a simplified form of reconstruction to be expected: a variation of the distance between atomic layers, notably between the first and second layer. As the crystal tends to assume a minimum of the total energy, the surface energy will be minimum in thermal equilibrium. From an intuitive analogy to the macroscopic surface tension one would expect a surface contraction, to minimize the free surface. Microscopically, however, this need not be the case. The total energy could also be minimized by a surface expansion in particular cases, e.g. by a reduction of the electronic energy due to reduced electron density (Bohnen /1981/, Rose and Dobson /1981/). Several previous LEED investigations found top layer contractions for W(001) in the range from 0 % to 12 % /Read and Russel, 1979/. An example for a theory-experiment comparison with polarized LEED is shown in Fig. 4.3. For the (1,1) beam at normal incidence ($\phi = 0°$) the reflec-

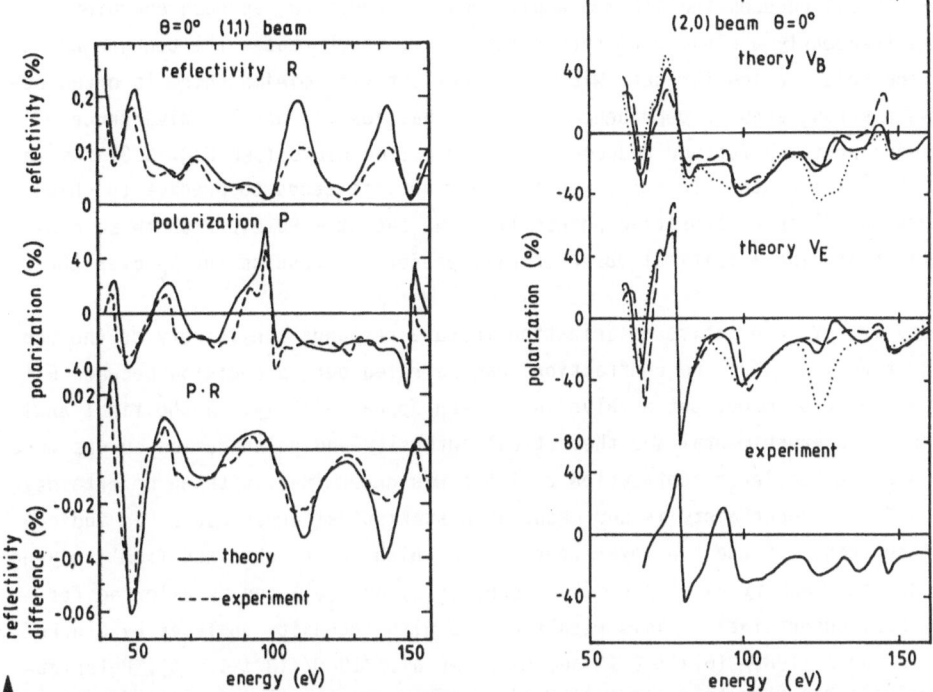

Fig. 4.3. Comparison of absolute values from theory and experiment for the (1,1) beam at normal incidence for reflectivity R, polarization P and reflectivity difference P·R. A surface contraction of 7 % is assumed in the calculations

Fig. 4.4. Theoretical polarization data for the (2,0) beam with two different potentials for three values of the surface contraction: 0 % (...), 5 % (——) and 10 % (---) compared to experiment

tivity R, the polarization P and the reflectivity difference P·R are given. The agreement between experiment and theory (assuming a contraction of 7 %) is quite good for all three criteria, for the polarization even somewhat better than for the intensities. Note, that the comparison is made between absolute quantities in theory and experiment. The question arises, how sensitively this agreement depends on changes of the electronic and structural parameters. For this purpose Fig. 4.4 is cited, comparing several theoretical polarization curves to the experimental one of the (2,0) beam at normal incidence. The two theoretical sets differ by the muffin-tin potentials. V_B corresponds to a non-relativistic, non-selfconsistent band-structure potential, V_E refers to a relativistic, selfconsistent potential with an energy dependent exchange part /Feder and Kirschner, 1981/. The potential V_B does not lead to a good agreement with the experiment, for all contraction values. Potential V_E brings a substantial improvement, in particular the structure between 80-100 eV is better reproduced.

At this point a general comment on the measurement of large polarization values seems appropriate. At energies where the polarization goes from strongly positive to strongly negative within a few tenth of an eV the energy width of the primary beam (~ 0.5 eV) reduces the maximum amplitudes. In addition, at such energies there is frequently a minimum of the reflectivity, which causes the background to reduce the polarization further. With sharp reflectivity minima there is often associated a strong angular dependence, which emphasizes the angular divergence of the primary beam and further reduces the polarization peaks (see Fig. 3.3). As in theory these influences are not taken into account, it tends to predict too high polarization values at sensitive points like the one at ~ 80 eV. With these considerations in mind a critical observer will prefer the results for V_E over those for V_B.

Around 125 eV, a negative polarization feature responds sensitively to the top layer contraction. Thus 'no contraction' can be ruled out, a decision between 5 % and 10 % cannot be made, but a value in between appears likely. In the final analysis about 100 experimental and theoretical intensity and polarization curves were included and a top layer contraction of 7.5 % was determined, with an uncertainty of ± 1.5 %. The uncertainty is not meant as a statistical error but as an indication of the range of the top layer contraction which can be obtained from studying different beams and types of parameter scans (P(E) or P(ϕ)) and by allowing for experimental uncertainties. This result agrees with intensity analyses by Clarke and de la Garza /1980/ (6.7 ± 2 %) and Marsh et al. /1980/ (8 ± 1.5 %). Polarization analysis has thus been shown to yield surface structural information with an accuracy at least comparable to that of conventional intensity analysis. In absolute terms, the top layer distance was found to be smaller by 0.011 nm than the bulk value (0.158 nm) with an error margin of about ± 0.002 nm.

The reconstructed W(001) surface at a temperature of 110 K, which shows a $(\sqrt{2}x\sqrt{2})R\ 45°$ superstructure was investigated by Wang et al /1982b/ using a polarized electron source. The asymmetries A(E) at low temperature were almost identical to those at 420 K, as were the intensities I(E). This does not mean, however, that the polarization is insensitive to the reconstruction. Rather, a search for more sensitive regions of the parameter space (E,ϕ,θ,g) appears appropriate.

Spin-polarized LEED of Au(110) was investigated by Müller /1979/. This surface exhibits a (2x1) superstructure at temperatures below 700 K. An example of the experimental results is shown in Fig. 4.5 for the (0,0) beam at E = 50 eV with the scattering plane normal to the chains of the reconstruction model. In Fig. 4.5a the scattering angle α was varied at different temperatures. At 320 K there is a strong polarization feature at α ≈ 96°, which decreases with increasing temperature. Another feature at α ≈ 76° grows with increasing temperature. The polarization data at these angles are shown as a function of temperature in Fig. 4.5b together with those of the (0,1/2) beam. Above 700 K strong variations appear, which are correlated with an increasing diffusivity of the superstructure beams. From these observations it was concluded that an order-disorder transition is present. The reconstruction model assumes that every second densely packed chain along [1Ī0] is missing on the top layer. With increasing temperature the chains may

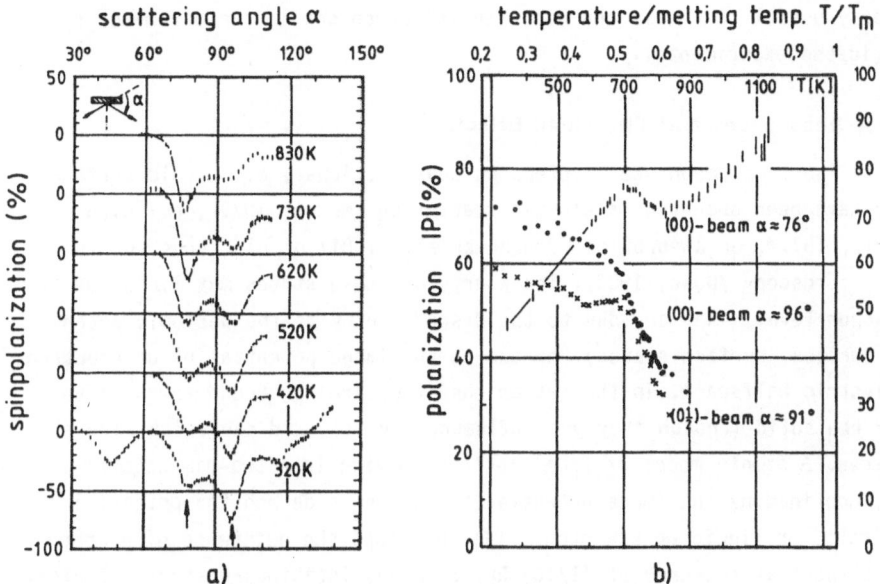

Fig. 4.5. a) Spinpolarization of the specularly reflected beam from the reconstructed Au(110) surface as a function of the scattering angle α at different temperatures (E_p = 50 eV, φ = 90°).
b) Magnitude of the polarization of the (00) beam at α ~ 76° and α ~ 96° and of
the (0,1/2) beam at α ≈ 91° as a function of temperature.
The order-disorder transition around 700 K leads to pronounced changes of the
spinpolarization

break up, shift by one atomic row, and may reduce their length until no sufficient
long range order exists any more. This model was essentially confirmed by recent
Scanning Tunneling Microscopy results /Binnig et al., 1983/. At high temperatures
the surface is thought to be formed by the unreconstructed substrate lattice, covered with an amorphous layer of short pieces of broken chains. A quantitative interpretation of the present data is not yet available, the phenomenon being too
complex for the present theoretical methods. A comparison of theoretical results
was made with data for the (1x1) surface (above 700 K). For a number of diffracted
beams reasonable agreement was found, for others not /Feder et al., 1977/. The origin of the discrepancies probably lies in the difference between the idealized
surface assumed in the calculations and the complex structure of the real surface.

Summarizing the first two sections of this chapter we may state: The general
symmetry relations between asymmetry and polarization on the one hand, and their
dependence on the symmetries of the crystal on the other hand have been veryfied
experimentally. The spin polarization is highly sensitive to structural parameters
and polarization analysis is a desirable complement to intensity analysis. For
relatively simple reconstructions, like a top layer contraction, fairly accurate
values are obtained. The analysis of more complicated structures depends on further developments on the theoretical side. Approximate but fast techniques would

be particularly useful to assess the most sensitive regions in the parameter space to be investigated experimentally.

4.1.3 Surface Resonances and Threshold Effects

The expression "surface resonances" refers to quasistationary electronic surface states. They have been observed in elastic scattering /McRae, 1979/, emission /Willis et al., 1977/, or absorption /Jonker et al., 1981/ of electrons, or in isochromat spectroscopy /Dose, 1983/. The energy of these states may lie below or above the vacuum level. They are due to the discontinuity of the periodic potential at the surface (surface states), or due to the image potential of an electron above a dielectric halfspace. In the latter case they are not bound to the crystallinity of the solid (though they are influenced by it), and can exist also at liquid surfaces. A simple model of free electrons moving in a two-dimensional 'waveguide', confined by the image potential on the one side and the potential step of the solid or liquid on the other side, predicts the existence of Rydberg states with a spectral constant of (1/16) Ry. Such quasistationary states of electrons above the surface of liquid helium have been investigated extensively by Grimes and co-workers /1978/. In this case the states are long-lived and there are electronic transitions observable between them. Because of the weak image potential (small polarizability of the He atoms, hence small dielectric constant) the characteristic transition frequencies are in the microwave range. On metals the potential is much stronger and the characteristic energies are of the order of 1 eV. Very long-lived states are not to be expected in general because the electronic density of states does not vanish in the energy range of these levels. The wavefunctions therefore may overlap with the 'tails' of the Bloch states extending into vacuum, which leads to strong damping of the image potential states. If there is a band gap in the surface Brillouin zone, however, the coupling is removed and the resonances become stationary and sharp. As the damping also depends on the spatial overlap, higher terms of the Rydberg series with a probability density farther away from the surface may become sharper and better observable than the lower terms.

The reason why we discuss surface resonances in the context of spin-polarized electron diffraction is that the LEED state, which we know to be spin-dependent, may couple to the image potential states under certain conditions. An incident electron may become 'trapped' in these states for a certain time, even at energies well above the vacuum level, which leads to intensity variations in the backdiffracted beams. The phenomenon is similar to resonance scattering in electron-atom scattering and its occurrence may be made plausible in the following way: At very low energies of the primary beam only the (0,0) beam exists. With increasing energy the radius of the Ewald sphere grows and eventually meets one or several 'rods' of the two-dimensional reciprocal lattice. At this energy the emergence threshold

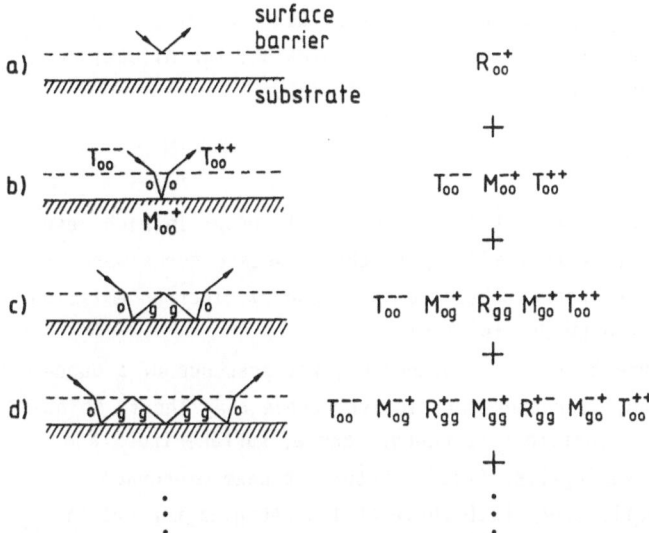

a) surface barrier / substrate

R_{oo}^{-+}

$+$

b) T_{oo}^{--} T_{oo}^{++} / M_{oo}^{-+}

$T_{oo}^{--}\ M_{oo}^{-+}\ T_{oo}^{++}$

$+$

c) $T_{oo}^{--}\ M_{og}^{-+}\ R_{gg}^{+-}\ M_{go}^{-+}\ T_{oo}^{++}$

$+$

d) $T_{oo}^{--}\ M_{og}^{-+}\ R_{gg}^{+-}\ M_{gg}^{-+}\ R_{gg}^{+-}\ M_{go}^{-+}\ T_{oo}^{++}$

$+$

$\vdots$ $\vdots$

Fig. 4.6. Schematic representation of direct (a+b) and indirect (c, d, ...) re-flection processes for the (0,0) beam at a metal surface with a surface barrier. The matrices R and T describe the reflection and transmission properties of the surface barrier, M represents the reflection at the substrate. In process c the (0,0) beam exchanges a reciprocal two-dimensional lattice vector g with the sub-strate, thus leading to interference with a pre-emergent beam

for one or several higher order beams is reached. Near threshold the beams run nearly parallel to the surface, the electron motion corresponds to that of elec-trons in image potential states, and they couple to the latter. This quasistation-ary state may decay either by inelastic processes, like the excitation of surface phonons (Rahman and Mills /1980/, Tong et al. /1981/), or by elastic diffraction. As the two-dimensional periodic potential acts on the electron it may exchange a reciprocal lattice vector with the crystal and be diffracted back into the (0,0) beam. In this beam (and other backdiffracted beams), intensity oscillations appear with the periodicity of the Rydberg series converging to the emergence threshold.

Below threshold the electron motion may be as visualized in Fig. 4.6d, being reflected back and forth several times between the surface barrier and the sub-strate until it finally escapes into the (0,0) beam /Le Bossé, 1981/. The ampli-tude of the specular beam is described by M. The number of orders to be taken into account is determined by the 'direct' contribution (terms a and b) as well as one or more of the 'indirect' contributions c, d,... . The transmission and reflection at the image potential is described by the matrices T and R, while the reflection at the substrate is determined by the damping: many reflections for weak damping, few for strong damping corresponding to efficient coupling to Bloch states (hence small reflection amplitudes). The contributions from the higher order terms are superimposed onto the 'normal' intensity variations resulting from the Bragg con-dition for Bloch states (cf. Sect. 2.4). They are small in general and are fre-quently only made visible by electronic or numeric differentiation /McRae, 1979/.

The existence of resonances induced by the surface barrier was confirmed by Rundgren and Malmström /1977a,b/ in nonrelativistic LEED calculations for Al(001). For the energetic level E_n of the resonances they gave the formula:

$$E_n = \frac{1}{16} (n+a_n)^{-2} \; [Ry] \; n = 1, \; 2 \; \ldots$$

where a_n corresponds to a 'quantum defect' that depends weakly on n. In high resolution measurements by Adnot and Carette /1977/ up to three Rydberg terms were observed. The relation to surface resonances observed in secondary electron emission /Willis et al., 1977/ was discussed by Willis /1981/.

The condition for the existence of surface resonances, the presence of a bandgap in the surface Brillouin zone, is rather restrictive. Hence they should be observed only under rather special conditions of energy, angle, surface index and material. In reality, however, Rydberg-like fine structure at beam emergence thresholds is observed in virtually every LEED curve at low energies for practically all materials.

It was pointed out by Le Bossé /1981/ that it is not necessary to invoke quasistationary surface states in the above sense to explain the phenomenon of fine structure in LEED intensity curves. He found that in the majority of cases it is due to a simple interference between the direct wave (case a + b in Fig. 4.6) and the one indirect (c) which undergoes only one internal reflection at the surface barrier. While resonances in the proper sense imply terms of order c, d ... only, the interference implies also the zero order term a + b. Unfortunately, there is no distinction to be made between these two processes as the interference also produces Rydberg series. Therefore it appears appropriate to attribute the fine structure at beam emergence thresholds to an interference in general, rather than to the excitation of a surface resonance, except for a few special cases. The fine structure in the (0,0) beam from W(001) could be one of these (Le Bossé /1981/, Le Bossé et al. /1982/).

In the above discussion the electron was treated as a 'spin-less' particle. For the image potential this approximation appears justified because the potential gradients involved are far too small to cause significant spin-orbit interaction. For the substrate barrier this assumption is not valid, certainly not for heavy materials but also not for a relatively light element such as Ni as we shall see in Chap. 5. In the picture of Fig. 4.6 the polarization effects arise in the reflection from the substrate, characterized by the matrices M. Away from beam emergence thresholds the polarization of the specular beam is due to process b. It is clear then that at emergence thresholds polarization structures should appear, either via interference with process c or via scattering into surface resonances. To say it in other words, in the quantum mechanical description the states in the image potential as well as the Bloch states have to be represented by (at least) two-component spinors. The matching of plane wave spinors in the vacuum, Bloch

90

spinors in the crystal and 'surface spinors' at the interface becomes spin-dependent if spin-orbit coupling is significant at some part in the system - in this case in the crystal. The existence of spin polarization effects at beam thresholds was predicted by numerical calculations for Cu(001) /Jennings, 1971/, Ni(111) /Feder, 1977a/ and W(001) (Feder /1974/, Jennings and Jones /1978/). As expected, the polarization features reacted very sensitively to the details of the assumed surface barrier, and also to an absorptive potential. For example, the assumption of a refractive but non-reflecting barrier (processes a and b in Fig. 4.6) makes the threshold effects vanish as the internal reflections (processes c, d ...) are absent. The existence of polarization effects was experimentally veryfied by McRae et al. /1981/ and calculated using a semi-empirical model on the basis of Fig. 4.6 (see Fig. 4.7). An analytical treatment of surface resonances with inclusion of the spin was carried out by Malmström and Rundgren /1981/. They found the qualitatively new result that under certain conditions, i.e. for interacting surface resonances, the Rydberg levels for ℓ emergent, non equivalent beams may split up into 2ℓ sublevels. The double-peak feature in the majority intensity (Fig. 4.7, $\Theta = 15°$) was interpreted accordingly, but Jones and Jennings /1983/ showed by numerical LEED calculations that this new type of resonance processes needs not to be involved to explain the experimental results. They demonstrated also that only spin-dependent calculations satisfactorily reproduce the observed intensity interference structure. These calculations provided data for the position and the depth of the surface barrier, a quantity that is of great importance in all low energy electron studies.

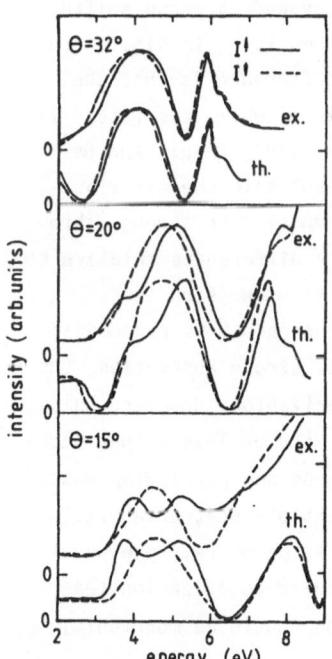

Fig. 4.7. Lineshapes and spin dependence of the reflectivity of the (0,0) beam from W(001) in the vicinity of the emergence threshold of the (0,1) beam for different polar angles of incidence ($\phi = 0$). Calculated (th) and experimental curves (ex) are normalized to each other. $I^{\uparrow}$ indicates the primary beam polarization being oriented along the normal to the scattering plane (after McRae et al. /1981/)

4.1.4 Temperature Effects

In Sect. 4.1.2 we had a case of temperature dependent polarization which was related to an order-disorder transition at the Au(110) surface. The polarization was taken as a sensitive monitor of this transition, but the effects of a temperature rise only, without a structural change were not discussed. For this purpose we choose a simpler system, the W(001) surface, well above room temperature, where it is not reconstructed.

We remember from Sect. 2.4. that in the kinematic approximation the polarization is the same as that of an atom. In this model the polarization should be strictly temperature-independent and there is a publication by Calvert et al. /1977/ reporting on the polarization of the (0,0) beam from W(001) being independent of temperature. Independent measurements by Kirschner and Feder /1981/ for the same nominal scattering conditions showed that temperature effects do indeed exist. They are relatively small for these particular conditions and the measurements have to be extended over a large temperature range. From this work it is evident, that a kinematical point of view is not appropriate when considering the temperature dependence of polarized LEED.

Raising the temperature we should expect the following to happen: At very high temperatures, close to the melting point, the vibration amplitudes of the crystal atoms become very large. As discussed in Sect. 2.4, the scattering amplitudes are weakened and the diffuse background increases. For a single vibrating atom this means that less flux is scattered than is incident on the ion core /Pendry, 1974/. This corresponds to absorption and therefore the scattering behaviour of the vibrating ion core may be described by complex temperature dependent phase shifts. Their effect - a reduction of the coherently scattered intensity - is similar to that of the absorptive part of the inner potential. The difference is that the former absorption effect is localized at the muffin-tin atom, while the latter is generally assumed constant (exceptions: Rasolt and Davis /1979/, Rasolt and Davis /1980/). The simplification of the Debye model assuming isotropic thermal vibrations was investigated by Ulehla and Davis /1978/. For a model with strong vibration amplitudes normal to the surface only small intensity differences relative to the isotropic case were found. For the spin polarization this question is yet open. In any case, the importance of multiple scattering processes is reduced at elevated temperatures. In the high temperature limit, i.e. strong absorption, the kinematic limit should be reached, with an atom-like polarization. However, this case corresponds more to scattering from liquids, with a more or less strong background reducing the polarization - to what extent depends on the particular case.

Because of the gradual transition from the dominant multiple scattering regime at low temperatures to kinematic-like scattering at high temperatures, one should in any case expect a change in size and shape of a particular polarization feature. One should not conclude, however, that the variation should be monotonic, and that the high temperature limit necessarily is characterized by reduced polarization. For example, it was noted by comparing polarization effects from solid and gaseous mercury /Eckstein, 1967,1970/ that the polarization may be large and very similar in both states. We remark that for a general scattering condition, where the polarization vector is not normal to the scattering plane at low temperatures, a temperature dependent rotation of the P vector in space should be observed because in the kinematic limit the polarization vector is always normal to the scattering plane. Such an effect could perhaps serve as an experimental key to answer the question at what temperature and momentum transfer the kinematic limit is reached.

At lower temperatures, in the multiple scattering regime, the temperature dependent modifications of multiple scattering occur in parallel with the thermal lattice expansion. At fixed primary energy the relation between electron wavelength and lattice constant changes. For the intensity this means - besides the decay according to the Debye-Waller factor - that the shape and energetic positions of peaks in a spectrum are modified as the electronic band structure varies with a variation of the lattice constant. In the simplest case of Bragg-like intensity peaks they are shifted to lower energy while the shape is approximately retained. In this approximation the thermal expansion coefficient of the lattice in the near-surface region may be measured, and this has been done in conventional LEED /Webb and Lagally, 1973/. The reader is referred to the review article by Lagally /1975/. It should be said, though, that because of multiple scattering there is frequently a change in the shape of the peaks, which makes an analysis relying on the kinematic approximation uncertain. The lattice expansion coefficient determined in this way always is an 'effective expansion coefficient', averaged over a certain number of atomic layers. It was found generally that the effective expansion coefficient is approximately equal to the bulk expansion coefficient with an uncertainty of up to ± 40 % due to the above-mentioned difficulties. What can be measured in a relatively precise way is the effective Debye temperature. Because of the exponential intensity decay with temperature, and if 'pathologic' scattering conditions are avoided (due to multiple scattering), the effective mean square vibration amplitudes may be estimated fairly well. By varying the primary energy and the scattering conditions the surface region may be scanned within certain limits, and the mean square amplitudes of the surface atoms be determined. For vibrations normal to the surface it was found that the mean square amplitude is two to five times larger than in the bulk for different materials. For vibrations parallel to the surface this value is more difficult to determine, but in general it was found to be much closer to the bulk value /Lagally, 1975/. Specifically for W(001) it was found theoretically by Black et al. /1980/ within the framework of the harmonic approximation, that at 400 K the mean square amplitudes of vibrations normal to the surface are larger than the bulk amplitudes by a factor of 2.15. The thermal expansion is a consequence of the anharmonic terms of the potential. At

the surface the anharmonicity is strong due to the missing neighbours, and in combination with the enhanced vibration amplitudes normal to the surface a temperature dependent expansion of the top layer distance (and eventually of the second
layer distance) is to be expected. At first sight this argument seems to be at
variance with the experimental finding quoted above that the effective surface expansion coefficient was found not to be very different from the bulk one. This can
be explained in the following way /Lagally, 1975/: The LEED measurements always
were made at 'kinematic' peaks. Therefore about 5 to 10 atomic layers coherently
add intensity, and, with increasing temperature, the layer distances increase
simultaneously for all layers except the first one. The relative contribution in
amplitude from the top layer therefore is small. In addition, its vibration amplitudes are large, further reducing the coherently scattered amplitude. Therefore
essentially the bulk expansion is measured. We arrive at the paradoxical situation, that in systems where the thermal surface expansion coefficient is large
(i.e. large vibration amplitudes) it cannot be measured because the intensity from
the top layer is too small. In cases where the intensity contribution from the top
layer is large (small vibration amplitudes) the surface expansion is too small to
be measured precisely. We shall see in the following that by means of polarization
analysis this paradoxon may be resolved.

An experimental and theoretical analysis was made by Kirschner and Feder /1981/
for the W(001) surface in the temperature range ~ 300 K to ~ 1200 K. Results for
the polarization of the (0,0) beam at $\Theta = 13°$, $\phi = 0°$ are reproduced in Fig. 4.8
for different temperatures. The polarization measurements P(T) where made at fixed

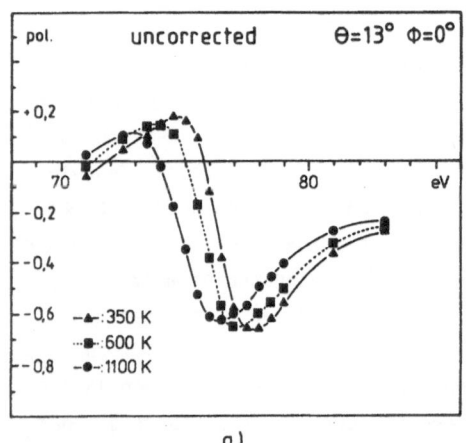
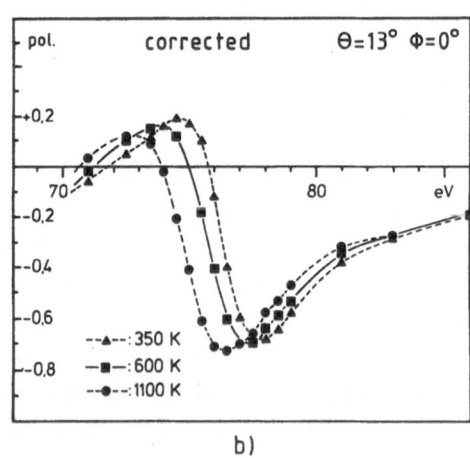

a)   b)

Fig. 4.8. a) Spin polarization of the (0,0) beam at $\Theta = 13°$ as a function of energy for three different temperatures. Note, (i) that the negative peak is relatively less affected than the positive one, (ii) that the feature shifts to lower energy with increasing temperature.
b) Spin polarization data of Fig. 4.8a corrected for the presence of the temperature dependent (essentially unpolarized) thermal background. Note that the positive peak still decreases with increasing temperature, whereas the negative peak
even shows a slight increase

energy while the crystal cooled down from elevated temperatures ('cooling curves'). Depending on energy, the polarization may increase or decrease or remain constant, in qualitative agreement with a previous observation by Riddle et al. /1978/ for nominally the same scattering parameters. From the cooling curves P(T) the energy dependent polarization curves P(E) were reconstructed for three temperatures (Fig. 4.8a). The thermal diffuse background was measured with respect to intensity and polarization and the polarization data have been corrected (Fig. 4.8b) by means of the formula

$$P = P'(1 + I_b/I) - P_b I_b/I$$

P' is the measured polarization, P and I are the polarization and intensity of the diffracted beam, and P_b and I_b are the polarization and intensity of the background. This correction is necessary as theoretical data do not take the background into account. In Fig. 4.8 we note the following: (i) with increasing temperature the polarization curves shift to lower energy; (ii) with increasing temperature the shape of the curves changes: the positive maximum is reduced while the negative maximum is slightly increased (in Fig. 4.8b). In the theoretical calculations the effects of lattice expansion and enlarged vibration amplitudes were studied independently. It was found that the shift is essentially due to the lattice expansion, while the shape changes are essentially caused by the decrease of multiple scattering. Simultaneous variation of both lead to a qualitatively correct description of the experimental result. A closer inspection, however, revealed the experimental shifts to be larger than the theoretical ones. The reason is, as found by varying the top layer distance in the calculation, that the temperature dependent expansion coefficient is larger at the surface than in the bulk by a factor of two to three. This value is considerably larger than the scatter in the results from intensity measurements. It is quite plausible, though, in view of the strong enhancement of the surface vibration amplitudes. Also, this value agrees well with the results of model calculations by Kenner and Allen /1973/ for (100) surfaces of bcc crystals using a 6-12 Lennard-Jones potential. One should note, that over the temperature range studied (up to about one third of the melting temperature) the top layer distance is still smaller than in the bulk. Therefore, what we really observe is the temperature dependent removal of the top layer contraction, which goes at a much higher rate than the bulk expansion.

Remains the question why spin-polarized LEED apparently is more sensitive than conventional LEED in studies of temperature dependent surface expansion. The answer probably is given by the observation that the strong polarization effects investigated occur at a pronounced minimum of the intensity. As its shape and energetic position change with temperature, the above arguments about the intensity maxima are reversed: If all layers below the first one generate an intensity minimum via destructive interference, the small out-of-phase contribution of the top layer contributes a relatively large effect which becomes accessible to measurement. In addition, the polarization is essentially the difference between two intensities, which may exhibit stronger sensitivity than the sum of two intensities. Spin-polarized LEED in this respect resembles a difference method with its inherently better sensitivity.

In summary, we note the following effects of crystal temperature on the spin polarization:

1) Increasing the temperature reduces multiple scattering and leads to changes of the size and shape of polarization features. The atomic (better: liquid-like) limit is reached at high temperatures, but a monotonic behaviour should not necessarily be expected.

2) At intermediate temperatures, where multiple scattering is significant, the spin polarization responds sensitively to the thermal lattice expansion as well as to the top layer distance.

4.1.5 Adsorbates

In the studies of clean surfaces we found the spin polarization to respond quite sensitively to details of the multiple scattering conditions, being determined by the surface geometry, by atomic scattering amplitudes, and by absorption effects. Therefore we would expect pronounced changes of polarization features upon absorption of foreign atoms or molecules, modifying the multiple scattering conditions, the structure of the substrate or the strength of spin-orbit coupling effects. Indeed, in some of the experimental studies very striking changes have been observed. Ordered adsorption on W(001) has been studied by Riddle et al. /1979/ (CO, c(2x2)), Mahan et al. /1980/ (N_2, c(2x2)), Wendelken and Kirschner /1981/ (O_2, p(4x1) and p(2x1)), and Wang et al. /1982b/ (H_2, c(2x2) and (1x1)). Disordered adsorption on W(001) was studied by Riddle et al. /1979/ for O_2 and CO. The system Ni(001),c(2x2)Te was investigated by Lang et al. /1982/.

The adsorption of H_2 on W(001) leads to a c(2x2) LEED pattern, as does the cooling of the clean surface below room temperature. While for the low temperature phase polarization effects were weak, for H_2 in some scattering geometries pronounced changes relative to the clean surface were found. It was concluded that the W substrate reconstructs upon H_2 adsorption and that the reconstruction is different for the two structures leading to the same c(2x2) pattern /Wang et al., 1982b/.

The disordered adsorption of CO and O_2 on W(001) /Riddle et al., 1979/ at room temperature was monitored by observing the (0,0) beam at $\Theta = 13°$ (nominal). At about 78 eV the clean surface produces a strong polarization feature (the same as in Fig. 4.8), which is gradually weakened upon exposures of 0.2 to 1.2 Langmuir (1 L = $0.75 \cdot 10^{-4}$ Pa·s). The amplitudes of all the polarization features decrease, but do not vanish. For strong polarization features this can partly be explained by an increase of the diffuse background which generally is only weakly polarized. The effects of O_2 and CO are qualitatively very similar.

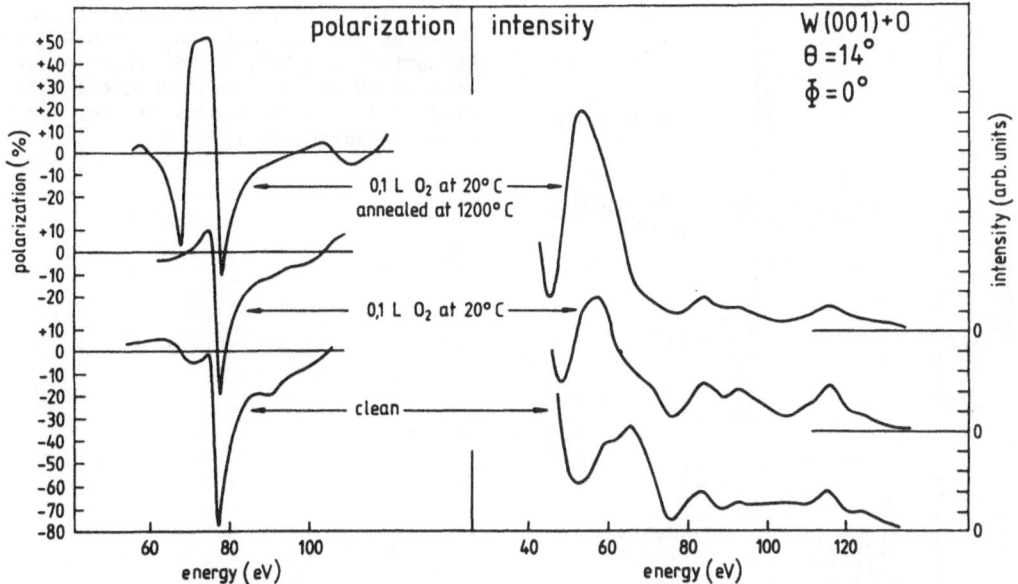

Fig. 4.9. The electron spinpolarization and intensity profiles for the specular beam from W(001) at $\Theta = 14°$ and $\phi = 0°$ after 0.1 Langmuir (= 10^{-7} Torr·s) exposure of oxygen and heat treatment

As an example for ordered adsorption the adsorption of O_2 is cited /Wendelken and Kirschner, 1981/. Fig. 4.9 shows the polarization and intensity of the (0,0) beam ($\Theta = 14°$, $\phi = 0°$) for small oxygen exposures before and after annealing. The changes of polarization and intensity after exposure of 0.1 L at ~ 20°C agree qualitatively and quantitatively with the above results of Riddle et al. /1979/ for disordered adsorption. Exposure of 0.1 L corresponds to a coverage of $\Theta_s = 0.1$ (relative to the number of W surface atoms) or less. This coverage did not modify the (1x1) LEED pattern. If the surface is annealed at 1200°C, the LEED pattern does not change but the polarization changes dramatically, exhibiting new oszillatory structures around 75 eV. It is tempting to attribute these changes to a reconstruction of the substrate, but it needs not be so. The adsorbate may change the surface barrier, modifying its transmissive and refractive properties. Also the multiple scattering between substrate and adsorbate layer may change the diffracted beam polarization. The weak changes upon adsorption at 20°C may perhaps be interpreted along this line, excluding spin-orbit effects from oxygen due to its small atomic number. On the other hand the dramatic effects of annealing, without the oxygen leaving the surface and without a change of the (1x1) LEED pattern, are hardly imaginable without a substrate reconstruction.

Comparing these observations with the results of Riddle et al. /1979/ and Mahan et al. /1980/ for ordered adsorption of CO and N_2, we observe striking similarities for all three adsorbates (see Fig. 4.10). The nominal polar angles are different by 1° from those of Wendelken and Kirschner /1981/ but a detailed analysis

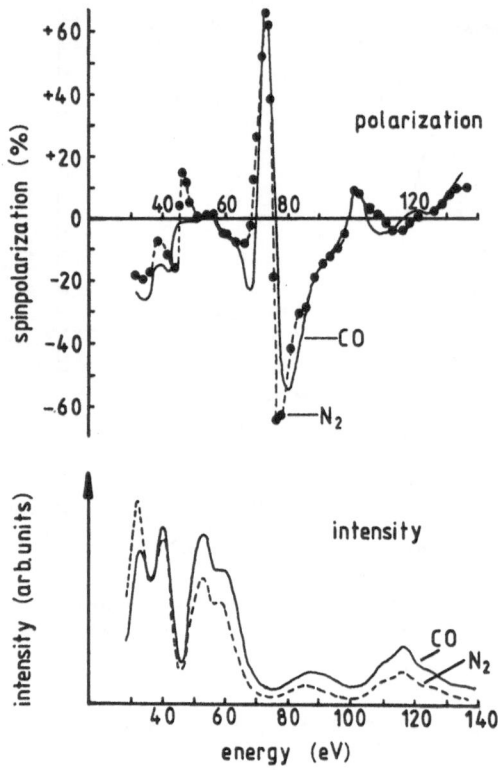

Fig. 4.10. Intensity and polarization profiles for the (0,0) beam from W(001) at nominal $\Theta = 13°$, $\phi = 0°$ after exposure to CO and N_2. Note the close correspondence to the curves in Fig. 4.9 after annealing at 1200° C

shows that the real angles agree to within 0.5°. Increasing exposure of N_2 at 250-350°C up to 1 L produces the polarization and intensities shown in Fig. 4.10. Superstructure beams appear diffuse at low intensity. Higher exposures change the polarization features, but annealing at ~ 1050°C brings them back while N_2 desorbs. A very similar polarization feature is generated after exposure of 20 L CO by annealing at 900°C until a coverage of $\Theta_s = 0.25$ is reached. These features are very similar to that in Fig. 4.9 after annealing, taking the small angular deviation into account. The coverages in all cases are much less than $\Theta_s = 0.5$, and we note that at the temperatures involved substantial surface diffusion may occur.

From the close similarity of the polarization features, independently of the existence of a long-range order, the conclusion must be that the W(001) develops the same or a very similar surface structure in all three cases, at least for low coverages. These systems should provide a suitable test ground for future dynamical polarized LEED calculations to determine the nature of the reconstructed surface, and the electronic properties of the top layer.

While most experimental studies were made with a light adsorbate on a heavy substrate, there is one with a heavy adsorbate (Te, Z = 52) on a Ni (Z = 28) substrate /Lang et al., 1982/. Even for the clean Ni(001) surface sizeable polarization was found, due to spin-orbit coupling. Upon absorption of Te in a c(2x2) structure new features appeared, which were generally stronger in polarization than for clean Ni, though less than P = 30 %. Initial theoretical calculations assumed the structure found in previous LEED studies with no reconstruction of the substrate. Reasonable general agreement between the calculated and experimental profile was observed. Some discrepancies indicated that further work is necessary to refine the theoretical model. The influence of multiple scattering between adsorbate and substrate was addressed by means of several model calculations. The polarization from a single two-dimensional array of Te atoms was found to be in disagreement with the experimental data and to be larger than that calculated for the Ni-Te system. This indicates that scattering from the Ni substrate is important and leads to reduced polarizations. Also, the polarization for a hypothetical c(2x2) Ni layer on a Ni(001) surface was computed. The maximal and average polarization were typically a factor of two smaller than those calculated for Ni(001),c(2x2) Te, indicating that the stronger spin-orbit coupling associated with Te does indeed lead to increased polarization.

At present, the number of experimental studies of adsorbate systems exceeds by far that of theoretical investigations. There is considerable promise, however, that the demonstrated sensitivity of the spin polarization to the presence of adsorbates may lead to improved structure analysis.

4.2 Spin-Polarized Photoemission

We recall from Sect. 2.5 that photoelectrons, even from non-magnetic crystals, will in general be spin-polarized. Considering the matrix element for photoemission (2.70) we have two possibilities for the occurrence of spin-dependent transition rates: (i) The final state wavefunction corresponds to the time-reversed LEED state, which may be spin dependent via spin-orbit coupling in the crystal. There are certain conditions where no polarization effects are observed, e.g. normal emission from crystals with inversion symmetry. (ii) Even for this case the photoelectrons may be found to be spin-polarized if the exciting light is elliptically polarized inside the crystal, as the selection rules for optical transitions may prefer a particular spin orientation. In this way, high polarization can be obtained even from relatively light materials like GaAs (see Chap. 3). With a three-step model of photoemission in mind we termed the first mechanism 'final state effect' and the second one 'operator effect'. We did this mainly for reasons of simplicity - in reality, both effects may occur simultaneously and may strongly interfere with each other. This may, e.g., be the case for photoemission from a heavy material with elliptically polarized light at a general direction of observation. We will maintain the distinction between the two effects in the following and will seek experimental situations where they appear in clean form.

4.2.1 Operator Effects

In almost all previous work it was the total yield of photoelectrons that was analyzed with respect to spin polarization, irrespective of energy and emission angle. For semiconductors we refer to the Sect. 3.2 on polarized electron sources
and the references given there. The high energy resolution experiment with GaAsP
reported there shows that the spin polarization is not necessarily constant over
the energy distribution, even when an integration over the emission angles is carried out. For metals, total yield studies were made by Zürcher et al. /1979/ on W,
by Heinzmann et al. /1972/ on Cs, by Pescia and Meier /1982/ and Borstel and
Wöhlecke /1983/ on Au, and by Humberg /1981/ on polycrystalline Cs. The latter
experiment was k̲-resolving in principle, but crystalline effects were largely
washed out as the photon beam averaged over many crystallites. The experiments
with Au and W were successful in determining band symmetries (e.g. the labeling of
the topmost bands of Au near the L point), but a detailed interpretation of the
polarization data (i.e. polarization of the total yield as a function of photon
energy) turned out to be fairly involved, due to the averaged nature of the experimental data. The reason why almost exclusively yield spectroscopies were made is
purely experimental: The intensity of conventional light sources in the ultraviolet range is fairly low. In combination with a Mott detector for spin analysis the
intensities obtained did not allow a further sacrifice of count rate such as to
reach substantial momentum resolution. It should be kept in mind that, unlike in
atomic physics experiments, the time for obtaining meaningful data from a clean
surface is severely limited, even in UHV. This can be circumvented by continuously
evaporating the metal onto a cooled substrate, thus maintaining a clean surface
(as Humberg /1981/ did), but with the disadvantage of not having a single crystal
surface. This situation changed when the first momentum- and spin-resolving photoemission study using a LEED detector demonstrated the feasibility of such experiments with discharge UV lamps /Kirschner et al., 1981/, and when synchrotron radiation became widely available. As is well known, synchrotron ratiation is linearly
polarized when viewed in the synchrotron plane and elliptically polarized outside
the plane. The handedness of the light depends on whether the charged particle
beam is viewed from above or from below the synchrotron plane. As the particles
generally are highly relativistic, at only a few mrad off the plane the radiation
is almost completely circular polarized ($P_{circ} \gtrsim 90$ %). Therefore, switching between left- and right-circular light does not require a movement of the monochromator or of its dispersing elements. It is sufficient to move an aperture at the
entrance slit, which lets the synchrotron light impinge either on the upper or the
lower half of the dispersing grid. /Heinzmann et al., 1982/. A particular problem
when monochromatizing circular polarized light by optical grids is the absorption
in the metal, leading to an ellipticity of the light and a rotation of the axes of
the ellipse in a wavelength dependent manner. These effects vanish for normal in-

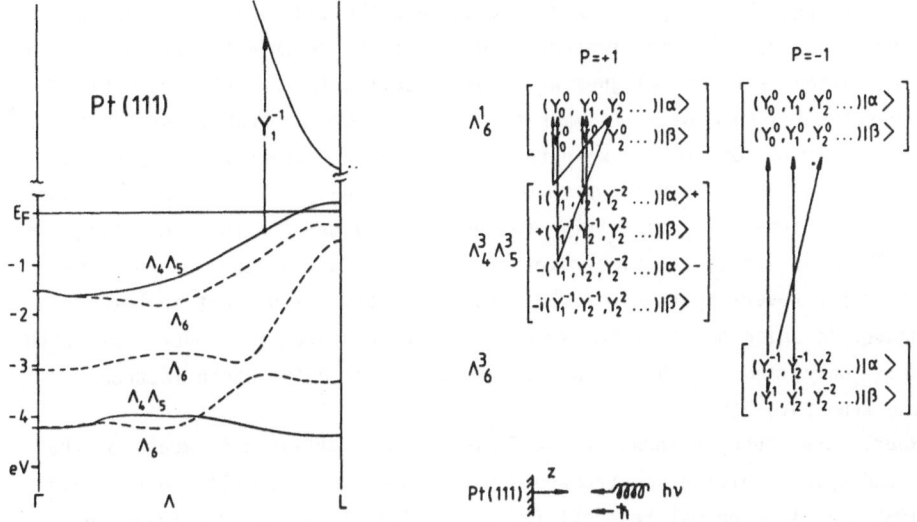

Fig. 4.11. Bulk bandstructure of Pt along the Λ line and corresponding Bloch states expressed as linear combinations of spherical harmonics, with the expansion coefficients suppressed. Application of the selection rules for circular polarized light represented by the operator Y_1^{-1} leads to plus or minus 100 % polarization of the electrons excited into the upper band of Λ_6^1 character

cidence and the monochromators therefore are of the normal incidence type /Eyers et al., 1983/. The availability of highly circular polarized light in the ultraviolet range at high intensities is a unique feature of synchrotron radiation.

Some first momentum- and spin-resolved photoemission experiments from solids with circular polarized UV light have recently been carried out at the german dedicated storage ring BESSY, Berlin (Oepen et al. /1983/, Eyers et al. /1984/). Pt(111) was chosen because this surface is unreconstructed (cf. Sect. 4.1.1), and because the spin-orbit splitting of the occupied states is sizeable according to previous experimental and theoretical investigations (Mills et al. /1980/, Andersen /1970/, MacDonald et al. /1981/). For emission from non-degenerate and non-hybridized bands a ± 100 % polarization was predicted from group theory by Wöhlecke and Borstel /1981a-d/ for the Λ line in cubic crystals. This result may be veryfied by a look at Fig. 4.11. The left hand part represents an extension of the RAPW (relativistic augmented plane wave) band structure of Andersen /1970/ to higher energies by Borstel (private communication /1983/). The right hand side represents the symmetry adapted wavefunction for the upper final band of Λ_6^1 symmetry and the two topmost occupied bands of $\Lambda_4^3\Lambda_5^3$ and Λ_6^3 symmetry respectively. The Bloch spinors are expressed as linear combinations of spherical harmonics Y_ℓ^m up to $\ell = 2$ with the expansion coefficients suppressed. The convention for the sign of the electron spin polarization and the light polarization is shown on the lower right. For positive helicity of the light the photon spin $\hbar$ is aligned with the photon momentum. In conventional light optics this corresponds to left-circular light. The light travels along the -z direction, and the light operator inside the crystal corresponds to Y_1^{-1}. The electrons are said to be positively spin-polarized if the polarization vector is aligned with the direction +z. For the transitions $\Lambda_6^3 \rightarrow \Lambda_6^1$ the selection rule $\Delta m = -1$ leads to excitation of $|\beta>$ states only, leading to a spin polarization of -100 %. We note that in the non-relativistic approxima-tion employed here for the dipole operator spin-flips $|\alpha> \rightarrow |\beta>$ or vice versa are not allowed. This scheme is completely analogous to the one for the Δ direction in Fig. 2.13. The $\Lambda_4\Lambda_5$ states are time-reversal degenerate and because of the time reversal operator reversing the spin, $|\alpha>$ and $|\beta>$ states are present in both spin-or components. In spite of this, the selection rule populates the upper state with $|\alpha>$ electrons only, and a +100 % polarization is predicted for photoelectrons originating from this band.

The experimental results shown in the following were obtained by means of the momentum- and spin-resolving spectrometer system shown schematically in Fig. 3.4 /Oepen, 1984/. Besides normal take-off from the (111) plane the experiment re-quired normally incident light, as otherwise the polarization state of the light would not have been known. The problems associated with the secondary electron production by the UV-light passing through the spectrometer and the electron transport lens were overcome by carefully collimating the incident beam and bi-asing the target. The angular resolution was then about ± 5° or less as determined by numerical ray tracing between the sample and the entrance aperture /Hünlich, 1984/. A typical result is shown in Fig. 4.12 for $h\nu$ = 13 eV. Intensity a and po-larization b are measured as a function of energy below the Fermi-level. The po-larization vector was found to be normal to the Pt(111) surface, so that the po-larization component in b equals the total length of the vector. The data for in-tensity and polarization are obtained simultaneously in one run. The total inten-sity may be decomposed into two partial intensities with polarization parallel or antiparallel to the surface normal. This is accomplished by means of the formulae $I_\uparrow = I_{tot}(1+P)/2$ and $I_\downarrow = I_{tot}(1-P)/2$. The result of this decomposition is dis-played in Fig. 4.12c. It is clearly seen that the leading peak close to the Fermi-level is positively polarized, while the second and third peaks are negatively polarized. A fourth weak peak has again positive polarization. In this way each peak in the total intensity curve can be classified according to the symmetry character of the band it stems from. In Fig. 4.12d the Pt band structure along Λ is reproduced, with the final Λ_6 band displaced in energy by $h\nu$ = 13 eV. Within the direct transition model between Bloch states a peak in the energy spectrum should be observed where the displaced final band intersects an initial band. This seems to be the case with very good accuracy. The bars in Fig. 4.12c represent the energetic position, the relative strengths, and the polarization sign of the al-lowed transitions as predicted by theory (Borstel, private communication).

From the sequence of the polarization signs it can unambiguously be concluded that the topmost band and the fifth band from the top must be of $\Lambda_4\Lambda_5$ symmetry,

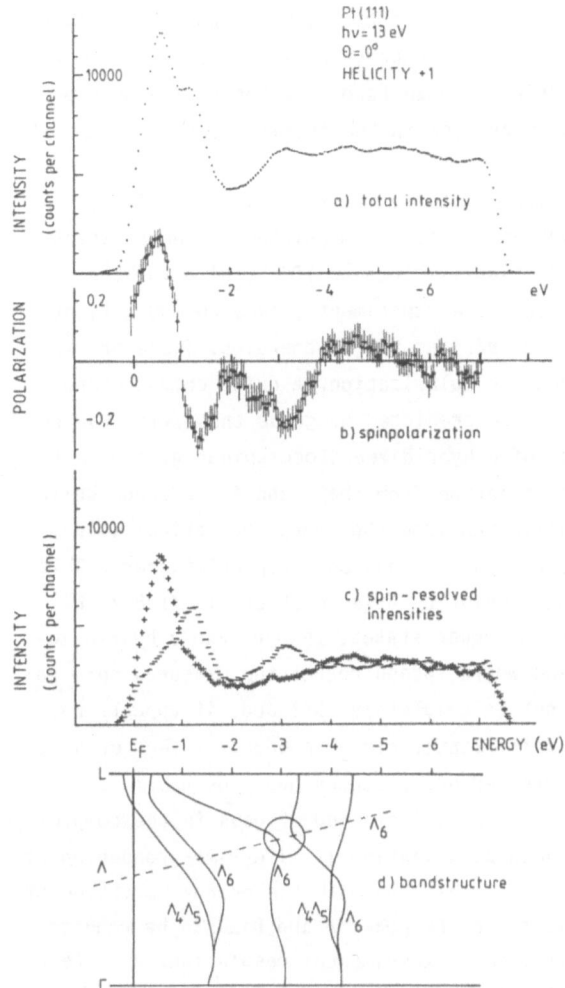

Fig. 4.12. Results for normal incidence, normal take-off photoemission from Pt(111) at hν = 13 eV with light of positive helicity. Measured are total intensity a) and spinpolarization b), from which the spin-resolved intensities c) may be calculated. The peak positions correspond to direct transitions in the bandstructure c) with the upper band displaced by hν

while the bands in between must be of Λ_6 symmetry. A point of particular interest is the encircled region where two bands come close to each other. In a previous band-structure calculation by MacDonald et al. /1981/ a bandcrossing is found, while otherwise the two theoretical band structures are in fair agreement. From the experimental data, however, it is evident that both bands lead to the same sign of the spin polarization. Hence the symmetry of the bands must be the same, and there must be an anticrossing of bands, instead of a crossing, since bands of like symmetry do hybridize. At this point the Bloch spinors of both bands contain states of Λ_6^1 and Λ_6^3 symmetry. In this way spin-polarized photoemission allows not only to determine the dispersion of bands but also their symmetry character. This additional degree of freedom we expect to be of great value in identifying electronic transitions and in labelling less well-known band structures. The only prerequisite is, that spin-orbit coupling is present in the initial state and that it can be resolved experimentally.

While we find good qualitative agreement with the theoretical predictions, there are quantitative discrepancies yet as far as the magnitude of the polarization is concerned. The theoretical prediction is close to 100 %, also for the more realistic calculations. When we extrapolate the experimental data to 100 % light polarization, we find only about ± 50 % at various photon-energies. There are several possible mechanisms leading to reduced polarization. A quite common cause for a measured polarization being other than predicted by group theory is hybridization. If the two (or more) substates of a hybridized Bloch spinor give rise to opposite polarization, the net spin polarization from that band is reduced. While hybridization may thus reduce the polarization from the group-theoretical value, it may also lead to a non-zero polarization when group theory predicts zero. This was nicely demonstrated for Ge near the Γ point by Allenspach et al. /1983/. Hybridization is of equal importance for the upper states. In our case a hybridization of the final Λ_6^1 band with a Λ_6^3 band would indeed reduce the measured spin polarization. While this possibility cannot be completely excluded, it appears unlikely as no band of this symmetry is sufficiently close in energy /Eyers et al., 1984/. In addition the damping of the excited upper states has a de-hybridizing effect. This effect is essentially the same as the closing of gaps in the complex band structure. A second cause for reduced polarization is life-time broadening of the initial states. If the broadening were comparable with the energy splitting of the two topmost bands the net polarization in the peaks A and B would be reduced. This effect is similar to that of insufficient experimental resolution. As a test, the photon energies were chosen such that transitions close to Λ were excited, at a k value where the $\Lambda_4\Lambda_5$ band passed across the Fermi-level. Though both disturbing effects were absent then, the measured polarization was even slightly smaller than at higher photon energies which rules out this possibility. A further effect that has already been mentioned, is the occurrence of spin-flips during the optical transition. Though the effect should be small in general, it might become sizeable under certain conditions, e.g. in d-band photoemission /Feuchtwang et al., 1978/. While the above discussion emphasized the primary excitation, there are also secondary effects leading to a reduction of the spin polarization. An experimental factor of considerable importance is the finite acceptance angle of the spectrometer. When observing outside a highly symmetrical direction the spin polarization is expected to decrease /Wöhlecke and Borstel, 1984/. Other measurements also showed, that the polarization vector is no more normal to the surface /Oepen, 1984/ when going off the Λ line. For these reasons a finite acceptance cone will tend to reduce the measured polarization (see also Borstel and Wöhlecke's

/1984/ analysis of the yield data from Au measured by Pescia and Meier /1982/).
Numerical estimates for the present case (Borstel, private communication) indicate
a reduction to 75 % from an initial 100 % polarization. A further effect is the
contribution from non-direct transitions. As already noted by Mills et al. /1980/
there is a rather strong density-of-states feature in Pt associated with a flat
band along the Q line, which gives a fairly strong intensity underlying the d-band
peaks. This density-of-states contribution presumably is unpolarized as it results
from non-k-conserving transitions. In the spin-resolved spectra of Fig. 4.12c such
an intensity common to both channels does indeed appear to be present. If this intensity were unpolarized, it should bring the polarization closer to the experimental value.

There are also specific surface effects to consider, notably surface umklapp
processes. By these effects polarized electrons emitted along other directions of
the crystal might be diffracted back into the surface normal, thus changing the
net polarization even if they are unpolarized. These surface effects as well as
final state effects can be treated quantitatively only within the framework of a
one-step theory of spin-polarized photoemission. Theoretical work along this line
is underway (Feder and Borstel, private communication). Spin-polarized photoemission then promises a detailed experimental insight into the symmetry properties of
electronic bands. Hybridization effects will become accessible to direct investigation in the whole momentum space and a deeper understanding of the photoemission
process from crystals may be reached.

As has been said several times it is the symmetry of the wavefunctions of a
particular band state that gives rise to the spin polarization of the photoelectrons. The symmetry properties are a consequence of the symmetry of the crystal,
i.e. of the long-range order of the array of atoms. It is this symmetry that makes
the wavefunction being composed of a particular set of spherical harmonics for
each spin state along a particular symmetry direction, as for example shown in
Fig. 4.11. Along another direction a different set of spherical harmonics would
appear. In other words it is the crystalline order that causes particular harmonics to be absent along particular directions and for a particular spin state,
which in turn causes the photoelectrons to be polarized. In a disordered system,
such as amorphous semiconductors or metals, the long range order is destroyed, only short range order, essentially confined to the next nearest neighbours, pre-
vails. Under these conditions the symmetry of the wavefunctions is lost, and the
missing spherical harmonics are mixed into each spin state because of the structural disorder. Though the selection rules for circular polarized light remain
valid, now both spin states may be excited simultaneously and the spin polarization of the photoelectron current vanishes. This sensitivity of the spin polarization to structural disorder has been nicely demonstrated by Meier and Pescia
/1984/ for Ge. For a single-crystal (001) surface near threshold a spin polarization of about 30 % was measured. An amorphous layer of Ge, produced by evapora-

tion, did not give rise to any polarization within the experimental error. Upon annealing to about 250°C amorphous Ge transforms into microcrystalline Ge, with crystal sizes of the order of 0.1 μm. The annealed Ge layer showed the same spin polarization as the single crystal surface! Apparently linear crystal dimensions of the order of 50 lattice constants are sufficient to fully establish the wavefunction symmetries, while short range order only does not. Perhaps the photoelectron spin polarization could serve as a measure of the degree of long range order in a partly disordered amorphous system, once a suitable calibration has been carried out.

4.2.2 Final State Effects

We recall from Sect. 2.5 that the final state wavefunction in the photoemission matrix element is identical to the LEED state. If the initial state is degenerate with respect to the electron spin, and if the dipole operator has no preference for a particular spin orientation, then the spin-orbit interaction in the LEED state may be the reason why spin-polarized electrons are found in the halfspace above the crystal. In the previous section we wished to observe 'operator effects' in the 'cleanest' possible way and therefore we chose the one singular point in space where final state effects vanish: at emission along the normal to the crystal surface. For centrosymmetric non-magnetic crystals the polarization induced by spin-orbit interaction must be zero in this case. In this section we want to study final state effects in the 'cleanest' possible way. Therefore we choose a non-magnetic crystal of high atomic number (W(001)). In order to suppress operator effects we can either use linear polarized light or unpolarized light. For linear polarized light there is no preference for a particular spin orientation, provided that the light is linear polarized inside the crystal. To assure this, one would have to use special incidence conditions, e.g. normal incidence. For unpolarized light this is not necessary, but normal incidence was used for experimental reasons. The UV light was obtained from a gas discharge lamp operating with D_2 for intensity reasons. Photoelectrons were analyzed at an angle of $\Theta = 70°$ with respect to the surface normal. This direction together with the surface normal defines the 'emission plane'. The spin- and momentum resolving spectrometer was similar to that shown in Fig. 3.4 except for the spin analyzer. Instead of the channel-plate system the two-channeltron-device of Fig. 3.1 was used. Therefore only the polarization component normal to the emission plane was measured. The resolution in energy and angle was 0.3 eV and ± 3° respectively /Kirschner et al., 1981/. Final state effects were observed in two modes for fixed photon energy (10.2 eV): scanning the energy of the photoelectrons at constant polar and azimuthal angles in the (110) plane, or scanning the azimuthal angle at fixed kinetic energy and polar angle (rotation diagram). These two modes overlap at $\phi = 45°$. The result of the energy scan mode is shown in Fig. 4.13, with experimental and theo-

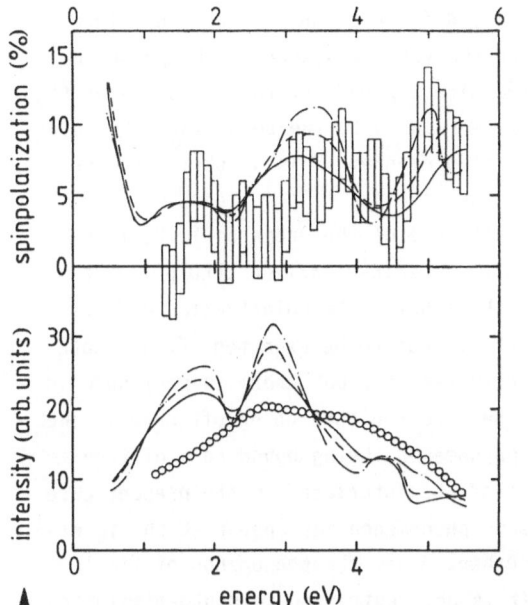
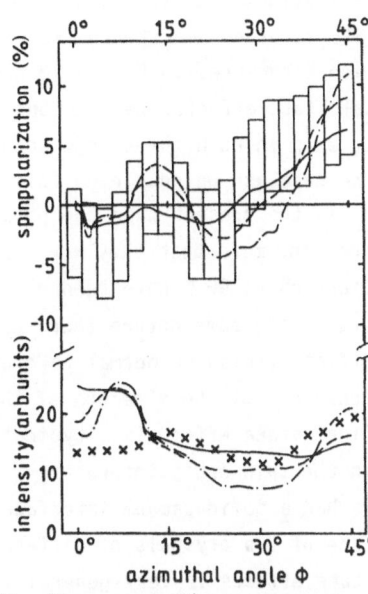

Fig. 4.13. Spinpolarization (upper panel) and intensity (lower panel) versus energy profiles of the photocurrent from W(001) for emission in the (110) plane at a polar angle $\Theta = 70°$ with respect to the surface normal. Theory for $V_i = 0$ (-•-•-), 0.25 eV (---) and (——) $V_i(E)$. The empty circles denote the experimental intensity

Fig. 4.14. Rotation diagrams of the component of the spin polarization vector normal to the emission plane (upper panel) and intensity (lower panel) of the photocurrent from W(001) for polar angle of emission $\Theta = 70°$ and kinetic energy $E = 5$ eV. Theory for $V_i = 0$ (-•-•-), 0.25 (---) and 0.67 eV (——). The experimental spin polarization is represented by rectangles, the height and width of which give the statistical error and, enhanced by about 20 %, the azimuthal angular uncertainty. The empty crosses represent the experimental intensity

retical results for the polarization in the upper part, for the intensity in the lower part. The vertical extension of the bars indicates the statistical 1σ error, the horizontal extension corresponds to the distance of the data points on the energy axis. The uncertainty of the polarization zero is ± 0.02. The results of the rotation diagram mode at a kinetic energy of 5 eV are shown in Fig. 4.14. In Figs. 4.13 and 4.14 there are three theoretical curves each for polarization and intensity for three different values of the damping, expressed by the imaginary part of the inner potential. As one might expect, the structures are washed out for strong damping, i.e. large V_{oi}. The theoretical data were obtained on the basis of the three-step model outlined in Sect. 2.5 /Feder and Kirschner, 1981b/. For simplicity, and because the emphasis was on final state effects, the matrix element for transitions between Bloch states was set constant. The intensity curves therefore refer only to the energy- and angle-dependent transmission through the surface. As the experiment measures also the effects from the matrix element a complete agreement of experimental and theoretical intensity curves is not expected and would be accidental. Accordingly, the agreement for the intensities is rather poor. The polarization, however, is a normalized intensity difference and the absolute intensity is of no importance. We emphasize, that the very existence of the polarization structures in Fig. 4.13 and 4.14 is the direct proof for the existence of final state effects. We remember, that if they were absent, the polarization should be zero in both the energy scan and the rotation diagram. This is obviously not the case and we note even fairly good agreement with the calculations.

In the present case the polarization effects are of the order of 10 %. This does not mean that they are small in any case. In spin-polarized LEED polarizations up to 90 % have been observed and as the final state polarization effects are of the same nature there are also larger effects to be expected. Final state effects vanish at normal take-off for symmetry reasons, but there is no monotonic behaviour of the strength of the effect to be expected when going off-normal. The final state effect is a typical interface phenomenon, being bound to a difference in the spin-orbit interaction at both sides of the interface. In the present case we had a solid-vacuum interface, but the same phenomenon may appear at the interface of two crystals of different atomic number. Since the phenomenon of final state effects is very general in nature, it is not restricted to photoemission but may appear in other emission spectroscopies as well, provided energy- and angle resolution is achieved. Final state effects are expected to be less pronounced for light materials, but it depends on the relative magnitude of initial state or operator effects to what extent they have to be taken into account. For example, in an initial state investigation, e.g. for Fe or Gd, they may lead to a change in magnitude and orientation of the polarization vector for off-normal emission, depending on the relative orientation of emission plane and magnetization direction.

Final state effects may also interfere with operator effects. They may lead to a rotation of the polarization vector or even to intensity asymmetries as pointed out by Kirschner et al. /1981/. At complementary angles with respect to the surface normal different intensities may be found if elliptical light impinges normally onto the surface. In analogy to LEED with a spin-polarized primary beam the spin-dependent matching of Bloch spinors to free electron states at the surface would lead to intensity differences at complementary observation angles /Kirschner et al., 1981/. Likewise, for a fixed take-off-angle the intensity spectrum may change when the handedness of the normally incident elliptical light is reversed. This effect is demonstrated in Fig. 4.15 for Pt(111). The intensity spectrum in the upper panel is characterized by two peaks, which correspond to emission from the same two upper bands as observed for normal take-off in Fig. 4.12. We see, that the relative intensities in these peaks change when the helicity of the light is reversed. The intensity asymmetry function is displayed in the lower panel of Fig. 4.15, which is reminiscent of the polarization curve in Fig. 4.12, except for the sign. The explanation of the effect is as follows: The take-off angle Θ is non-zero and the emission plane, formed by the surface normal and the momentum of the photoelectrons analyzed, includes a non-zero angle $\phi = 12.75°$ with the mirror

108

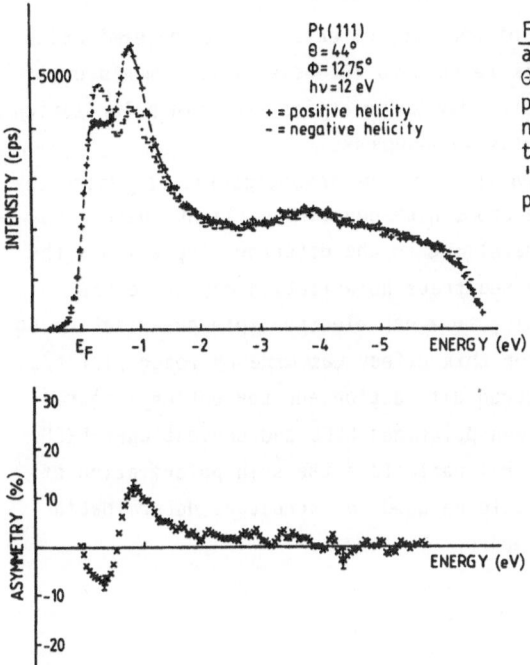

Fig. 4.15. Intensities (upper panel) and asymmetry (lower panel) for emission at Θ = 44° and φ = 12.75° off the ΓLUX mirror plane of Pt(111) for light of positive and negative helicity at hν = 12 eV. The intensity differences are a consequence of 'final state effects' in spinpolarized photoemission

plane of the crystal. Under these conditions the polarization vector inside the crystal has a component normal to the emission plane. We know from the preceding results from W that the transmission step through the surface acts as a polarizer for unpolarized electrons. We also know from the discussion of the double diffraction experiment in Sect. 3.1 that a polarizer can also be used as an analyzer, which, when based on spin-orbit interaction, gives rise to intensity asymmetries if the incident beam is polarized and if the vectors of polarization and asymmetry are collinear or have a collinear component. The first two peaks in the intensity spectrum have opposite polarization. Thus, for e.g. positive helicity, the intensity in the first peak is reduced, while that in the second peak is enhanced. Reversal of the light helicity reverses the initial polarization, which causes the first peak to be enhanced and the second one to be reduced in intensity. Thus the polarization of the photoelectrons inside the crystal is transformed into an intensity asymmetry outside the crystal. We note that the intensity asymmetry vanishes for φ = 0°, i.e. when the emission plane coincides with a mirror plane of the crystal. The asymmetry vector associated with the transmission step then stands exactly normal to the emission plane, but the initial polarization vector lies within the emission plane and has no component normal to it. Therefore the asymmetry vanishes (cf. (2.53)). This effect has in fact been used to locate precisely the mirror planes of the crystal relative to the emission plane. Once final state effects are known quantatively from experiment or theory, they could even be used for the quantitative investigation of operator effects, as the determination of the polarization inside the crystal is reduced to intensity measurements outside the crystal. In this sense the crystal may serve as its own spin polarization analyzer. Work along this line is currently in progress.

Final state effects could also be harnessed for the investigation of adsorbate systems. In photoemission from adsorbate atoms with unpolarized light there are interference effects between the wave travelling to the detector directly and the one reflected from the substrate. At the substrate polarization may arise via spin-orbit or exchange interaction, leading to a net electron spin polarization at the detector. A theoretical prediction for this effect was made by Feder /1977b/. The relation between polarized photoelectron diffraction and conventional photoelectron diffraction is the same as between polarized LEED and conventional LEED. As multiple scattering depends on structural parameters the spin polarization effects in photoemission from adsorbates could be used for structure determination, with the electron spin as an additional degree of freedom.

5. Results from Magnetic Materials

In this chapter we present and discuss experimental results obtained from magnetic materials. First, we treat elastic and inelastic electron scattering from crystals and amorphous metals (Sect. 5.1). In Sect. 5.2 electron emission spectroscopies are presented, by means of which the exchange-split electronic structure of ferro-magnets is investigated in detail. The unoccupied part of the electronic system can be studied by spin-polarized isochromat spectroscopy, which is presented in Sect. 5.3.

5.1 Electron Scattering

A classical example for the importance of the electron spin in scattering from magnetic surfaces was given by Palmberg et al. /1968/ in an unpolarized low energy electron diffraction experiment with NiO, which later was extended by Suzuki et al. /1971/. In addition to the normal diffraction spots, with reciprocal lattice vectors corresponding to the chemical unit cell, faint half-order spots were ob-served when the crystal temperature was lowered below a certain temperature. This temperature was found to be identical with the Néel temperature of the antiferro-magnetic NiO crystal. The appearance of the extra spots therefore is due to the formation of a magnetic unit cell of twice the dimensions of the chemical unit cell at low temperatures. The scattered amplitude depends both on the direct am-plitude f and the exchange amplitude g_{ex}. The latter one depends on the orienta-tion of the localized spins on the Ni^{++} ions. As neighbouring Ni^{++} ions have anti-parallel spins we have a periodic scattering potential with a periodicity twice the distance between neighbouring Ni^{++} ions. This is all that is needed to find a diffraction pattern with half-size reciprocal lattice vectors. This is one example where a magnetic information is coded in the intensity, even without explicit spin analysis. A second possibility is discussed below. In general, however, an analy-sis of magnetic properties will require an explicit consideration of the electron spin.

5.1.1 Elastic Scattering

We know from Chap. 2 that the vector quantities asymmetry $\underline{A}$ and polarization $\underline{P}$ are closely related to each other by time reversal symmetry and crystal symmetries. Elastic scattering experiments therefore can be done in two equivalent ways: either an unpolarized primary beam is used, followed by polarization analysis, or a polarized primary beam is used, followed by intensity analysis of the scattered electrons. Most experiments so far have been made with polarized primary electrons. The reason is that the effects to be studied are often rather subtle, requiring good statistics, i.e. high intensities. As the GaAs sources provide similar or even better intensities than unpolarized thermal electron sources, one is inclined to avoid the intensity loss of roughly three orders of magnitude encountered in even the most efficient spin polarization detector to date.

In the 3d ferromagnets spin-orbit coupling is relatively small, though by no means negligible. It turns out that exchange effects and spin-orbit effects are of about the same magnitude in Fe and Ni. For a general scattering geometry the interference of both effects may preclude a separate analysis of exchange effects. There is, however, the possibility of exploiting the symmetry properties of the crystal. An important simplification, often used in practice, comes about by the existence of mirror planes in the crystal. If the scattering plane coincides with a mirror plane, the polarization due to spin-orbit coupling is normal to that plane if unpolarized electrons are used. With polarized primaries an approximate decoupling of spin-orbit and exchange effects can be obtained if the polarization vector is normal to the mirror plane and parallel to the magnetization direction (cf. Sect. 2.4.2, Fig. 2.8a). In one asymmetry curve the sum of both effects is measured (A^+). If the magnetization is reversed, the spin-orbit contribution remains the same while the exchange contribution changes sign, resulting in the asymmetry curve A^- (cf. (2.59)). The exchange asymmetry A_{ex} may then be obtained by taking the difference to the variable baseline represented by the spin-orbit asymmetry A_{so}. This is illustrated in Fig. 5.1. (Alvarado and Weller /1984/, unpublished data). Fig. 5.1a shows the two experimental runs A^+ and A^-, obtained by reversing the magnetization. The exchange asymmetry A_{ex} and the spin-orbit asymmetry A_{so} are approximately obtained from A^+ and A^- by means of the formulae (2.60)

$$A_{ex} \approx \frac{1}{2}(A^+ - A^-); \qquad A_{so} \approx \frac{1}{2}(A^+ + A^-) \ .$$

It is evident, that as A_{ex} changes sign upon magnetization reversal, the arithmetic average of A^+ and A^- just gives the spin-orbit contribution A_{so} which is independent of the magnetization. The results of this decomposition are shown in Fig. 5.1b and 5.1c. This scheme works quite well for Ni and Fe as the spin-orbit coupling is weak in these materials.

As mentioned in Sect. 2.4.2 the interference between spin-orbit and exchange interaction may give rise to intensity changes of a particular diffracted beam

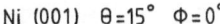

Ni (001) θ=15° Φ=0°

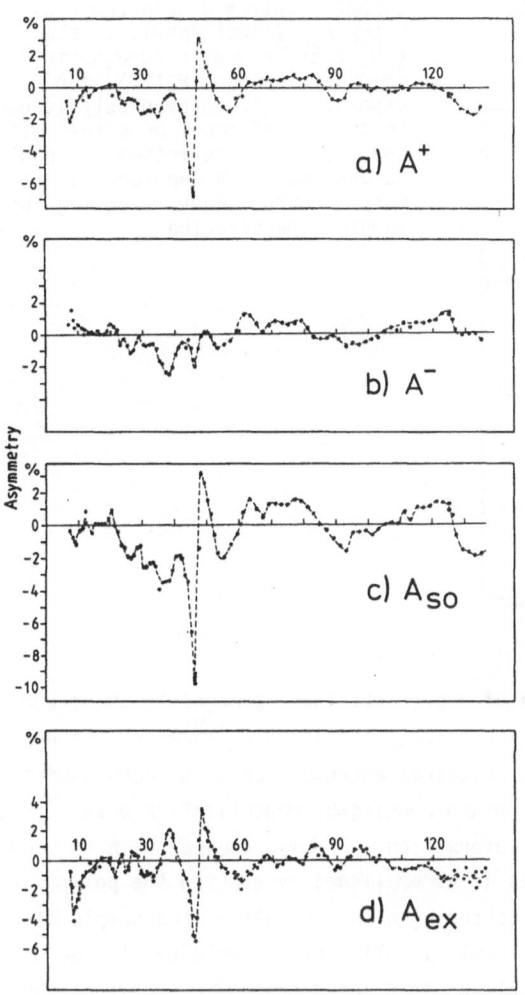

Fig. 5.1. Measured asymmetries A^+ and A^- for the (0,0) beam from Ni(001), obtained by reversing the sample magnetization. From these the spin-orbit asymmetry A_{so} and the exchange asymmetry A_{ex} are obtained as explained in the text

when the magnetization is reversed, even if the primary beam is unpolarized. While for 3d ferromagnets these effects appear to be very small, for heavier materials like Gd they become sizeable. This has been demonstrated experimentally by Alvarado and Weller /1984/ (see Fig. 5.2). The experiments were performed on a Gd(0001) layer (50 nm thick) grown epitaxially on W(110). Care was taken to eliminiate stray fields, which is much alleviated by using a thin film of magnetic material /Waller and Gradman, 1982/. The upper panel of Fig. 5.2 shows the experimental exchange asymmetry A_{ex}, the lower panel shows the spin-orbit asymmetry A_{so} in the specular beam as a function of energy. Both are seen to be of similar magnitude. The center panel displays the asymmetry A_u, i.e. the interference between both. As expected, it follows approximately the trend of the spin-orbit asymmetry A_{so}. Its

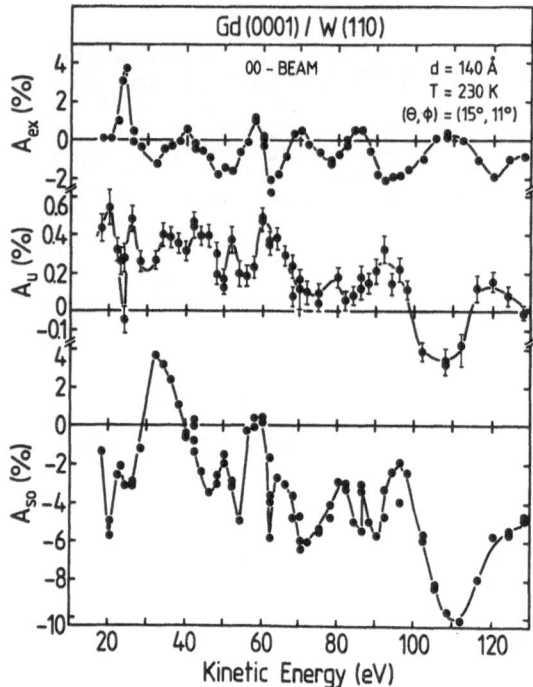

Fig. 5.2. Exchange asymmetry A_{ex} (upper panel) and spin-orbit asymmetry A_{so} (lower panel) in the (00) beam from a ferromagnetic Gd(0001) layer on W(110). The asymmetry A_u (center panel) is due to the interference of spin-orbit and exchange interaction. It can be observed with unpolarized primary electrons when reversing the sample magnetization

magnitude is of the order of a few tenth of a percent. Thus, a magnetic structure analysis could be done with unpolarized electrons, in principle. However, the nature of A_u, being an interference effect, requires extreme precision, both experimentally and theoretically, in order to obtain meaningful magnetization data.

If one is interested in the magnetic information A_{ex} only, one may wish to suppress spin-orbit effects altogether. This is accomplished by placing the polarization vector of the primary beam in the scattering plane as well as the sample magnetization ("planar geometry" cf. Fig. 2.8.b). Ideally, both should be aligned as then spin-orbit effects are exactly cancelled. If $\underline{P}$ has a component normal to the magnetization, the possible rotation of the polarization of the primary beam in the scattering plane (caused by the spin-flip amplitude g_{so} due to spin-orbit coupling) may lead to a slight additional intensity asymmetry. The experimental experience so far /Kirschner, 1984/ indicates, however, that spin-orbit effects are quite effectively suppressed in this scheme.

A quantity of essential interest when studying ferromagnets is the magnetization in the near-surface region. The magnetization in the first few layers of a solid can be probed well by electrons because of their short inelastic mean free path. A few studies have been made with clean single crystals of Ni (Alvarado et al. /1982a/, Feder et al. /1983a/) and Fe (Gradmann et al. /1983/, Tamura and Feder /1982/). The experimental exchange asymmetry curves $A_{ex}(\phi)$ or $A_{ex}(E)$ were compared to theoretical curves obtained under different assumptions for the magnetization in the near-surface region. As a common result the 'dead layer model', i.e. zero magnetization in the top layer, could be ruled out. To the contrary, there is evidence that the top-layer magnetization is larger than the bulk magnetization. For Ni(001) the top layer magnetization M_1 was found to be larger by 5 % ($\pm$ 5 %) than the bulk magnetization M_b if extrapolated back to zero temperature /Feder et al., 1983a/. Even larger effects were found for Fe(110). An example is shown in Fig. 5.3. (after Gradmann et al. /1983/). On the left hand side (a) and (c) the theoretical results with $M_1 = M_b$ are compared to experimental exchange asymmetry curves for two different angles of incidence ($\Theta = 30°$ and $\Theta = 45°$) in the (100) plane. The authors note good qualitative agreement for nearly all features. The most prominent exception appears at $\Theta = 45°$ near 70 eV and 75 eV, where a theoretical maximum appears near an experimental minimum and vice versa. Best agreement upon varying the relative surface magnetization M_1 was found for $M_1 = 1.3 M_b$, shown in Fig. 5.3 (b) and (d). The agreement is improved substantially, indicating an enhancement of the top layer magnetization. This finding is in general agreement with recent theoretical results from band structure calculations for

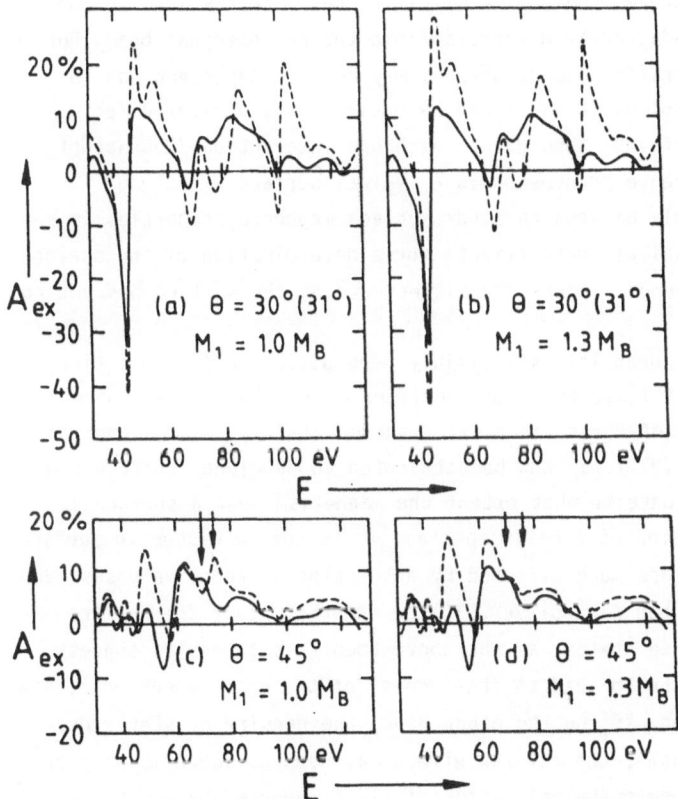

Fig. 5.3. Exchange asymmetry A_{ex} as a function of kinetic energy E for specular reflection of spin-polarized electrons on Fe(110). Experiment (full curve) for $\Theta = 31°$ and 45°, theory (dashed curves) for $\Theta = 30°$ and 45°. The top layer magnetization M_1 assumed in the calculations is expressed in units of the bulk magnetization M_B

thin slabs of Fe and Ni (Ohnishi et al. /1983/, Krakauer et al. /1983/, Jepsen et al. /1982/, Wimmer et al. /1981/, Wang and Freeman /1981/, Noffke and Fritsche /1981/, Freeman /1983/). It is mainly at the top layer where the magnetization is enhanced, being due to a large density of minority surface states near the Fermi level. According to the Stoner theory, which successfully predicts the presence of ferromagnetism for Fe, Co, and Ni and its absence for e.g. Cu, the density of states at the Fermi level is of crucial importance for the existence and magnitude of ferromagnetic order. The enhanced surface magnetization well below the Curie-temperature is thus consistent with Stoner theory. It should be noted that for very thin films (say five layers or less) the two surfaces may no more be decoupled from each other. Ferromagnetism is still present, but the magnetization may show an oscillatory behaviour with depth and/or film thickness (Wang and Freeman /1981/, Ohnishi et al. /1983/).

There is another phenomenon that may be harnessed to study surface magnetic properties. We know from Chap. 4 that certain fine structure in LEED intensity curves is related to the emergence of higher order beams at certain energies. This is an interference effect between the amplitude diffracted directly into the observed LEED beam and that temporarily diffracted into the pre-emergent beam. For non-magnetic surfaces the interference structures may be spin-dependent due to spin-orbit coupling in the solid. It was found recently that a similar effect exists also for magnetic surfaces, then due to exchange interaction /Rebenstorff et al., 1984b/. The interference phenomenon is a typical surface effect and its spin dependence could possibly be used to study surface magnetic properties in an alternative way. The potential of these effects for a determination of the surface magnetization or of spin-dependent charge densities outside the surface has yet to be explored.

The question of 'magnetic dead layers', vividly been discussed for some time, appears to be settled now, at least for clean surfaces of Fe, Co and Ni: there are no dead layers (for a review of the subject see Gradmann /1977/). Though the early findings /Lieberman et al., 1970/ may now be attributed to 'unclean' surface conditions, it is not clear to date to what extent the magnetism near a surface is influenced by the chemisorption of foreign species. It is common wisdom in surface science that surface states are much affected by adsorption of reactive gases even in quantities of much less than a monolayer. If the enhancement of the surface magnetism is caused by surface states, as the above theoretical results suggest, saturation of these states should lead to the removal of the enhancement or even a lowered surface magnetization. If, on the other hand, the density of states near E_F is enhanced by the adsorbate, the reverse effect may happen. There is also an extensive discussion on 'magnetocatalytic effects' first reported by Hedvall et al. /1934/ which means that the reaction rate of a catalytic reaction at a surface depends on the state of magnetic order of the solid. The first condition for such an effect to exist is the surface magnetization not being quenched by the adsorbate. Recent experiments on a basic surface reaction, the dissociative hydrogen adsorption on Ni surfaces by Robota et al. /1984/, show the reaction rate not to be affected by the breakdown of ferrogmagnetic order at the Curie temperature (for surfaces not contaminated by carbon segregation). Though there is apparently a need for investigations of the magnetism of surfaces covered by adsorbates, not very much has yet been done. The non-reconstructed Cr(100) surface was investi- gated by Meier et al. /1982a/ by means of photoyield spectroscopy. The clean sur- face emitted unpolarized electrons only since bulk Cr is antiferromagnetic. Incor- poration of a small amount of oxygen, most likely in a subsurface position gave rise to a spin polarization of up to 10 %, depending on the oxygen concentration (up to ~ 3 at % as determined by Auger spectroscopy). At first sight this result might intuitively be expected as chromiumdioxide (CrO_2) is known to be ferromag- netic. However, at least on the average, the O concentration was much too small to form this compound. One possibility to explain the observed magnetic ordering is that oxygen causes a redistribution of charge such that local magnetic moments are formed. Also, the incorporation of oxygen could distort the lattice, thus changing the sign of the magnetic interaction. This could occur also for small lattice changes as the exchange integral depends sensitively on distance /Siegmann et al., 1984/. A hint for a possible demagnetization of ferromagnetic surfaces by hydrogen absorption was given by field emission results from Ni (Müller /1975/, Landolt and Campagna /1977/, Landolt et al. /1978/). The spin polarization of field-emitted electrons from H_2-saturated tips in 10^{-8} Torr H_2 was observed to be zero within the experimental uncertainty. This was interpreted as being indicative of a demag- netization of the Ni surface at full hydrogen coverage /Landolt et al., 1978/. A different approach was recently used by Kirschner /1984c/, comparing exchange asymmetry spectra from clean Fe(110) with those from surfaces covered by O and S

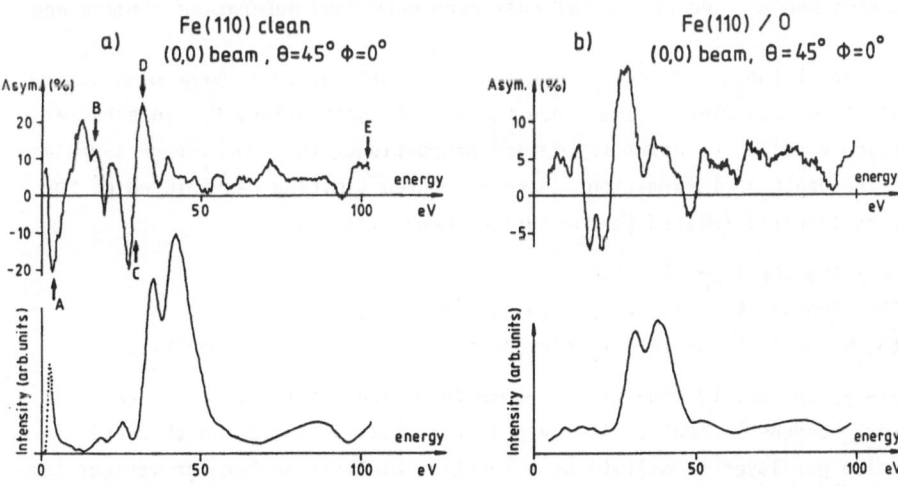

Fig. 5.4. Asymmetry and intensity profiles for the specular beam from Fe(110). a) clean surface, b) with ordered p(2x2) oxygen overlayer, corresponding to a quarter of a monolayer

in small amounts (about 1/4 of a monolayer). An example is shown in Fig. 5.4 for the (0,0) beam at an angle of incidence of $\Theta = 45°$. The intensity curve from the clean surface (Fig. 5.4a) is dominated by a Bragg peak at 40 eV, which is also the dominant feature for the oxygen covered surface (Fig. 5.4b). The structure of the intensity curves is very similar, except at very low energies where the surface barrier has large influence. There seems to be a constant background underlying the I-V curve for Fe(110) with oxygen, which could be due to diffuse scattering. The asymmetry curves refer to the exchange asymmetry only, as A_{so} is effectively suppressed by scattering in a mirror plane and using longitudinally polarized electrons. The structure of the asymmetry curves for the two surface conditions also is very similar, except in the low energy region. For the oxygen covered surface the magnitude of the most prominent features is reduced, while the asymmetry in the range from 50 to 100 eV is not changed very much. These effects could be due to the (presumably) unpolarized background. The depth of information in this energy range is about 2 atomic layers for the present geometry, and apparently the surface does not become 'magnetically dead' on this scale. Quantitative conclusions on the basis of exchange asymmetry data have to be drawn with due care as this quantity is not related to the surface magnetization in a simple way. For example, in Fig. 5.3c,d we found a polarization feature to be reduced in magnitude with increased top layer magnetization. Such observations are frequently made on isolated features of an asymmetry curve. On the whole, however, the asymmetry is larger for larger magnetization, as may be seen when comparing Fig. 5.3a,b and c,d. Exchange asymmetries generally are found to be larger for Fe than for Ni (magnetic moment 2.1 μ_B per atom in Fe versus 0.55 μ_B in Ni) and they do of course vanish for literally 'dead layers'. Quantitative conclusions can only be made in comparison to theoretical calculations, taking into account possible modifications of the surface barrier and structural rearrangements upon adsorption /Tamura and Feder, 1984/.

The relation between exchange asymmetry and magnetization is very complicated in general /Feder and Pleyer, 1982/ due to multiple scattering. For an arbitrary magnetization profile the asymmetry is *not* proportional to some average magnetization over the depth of information. There are three limiting cases in which the asymmetry is linearly related to the surface magnetization:

  i) if only the top layer is sampled,
  ii) if the magnetization is homogeneous in depth or,
  iii) if the magnetizations in all relevant layers scale by a common factor.

Unfortunately, in reality none of the above conditions is fulfilled in general. An idea of what happens in reality is given in Fig. 5.5. It shows the theoretical magnetization per layer of Ni(110) as a function of layer number for various temperatures from $T = 0$ to $T = T_c$ (t: $= T/T_c$), normalized to the bulk magnetization at $T = 0$ K. This calculation by S.-W. Wang /1980/ is based on a Heisenberg model

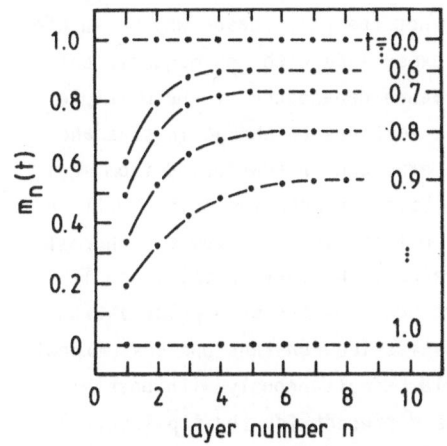

Fig. 5.5. Temperature dependent layer magnetization $m_n(t)$ ($t = T/T_c$) as a function of layer number as obtained form a Heisenberg-type model. Note that the magnetization decays faster at the surface than in the bulk and that the profile shape changes with temperature

assuming a layer-independent exchange coupling constant. The top layer magnetization enhancement of 13 % at T = 0, predicted by Krakauer et al. /1983/, is neglected here. We see, that the surface magnetization decays much faster with temperature than the bulk magnetization (compare layer 1 with layer 8), and that the shape of the profiles changes with temperature. At finite temperatures there is thus only one instance where we should expect a linear relationship between A_{ex} and the surface magnetization: at the Curie temperature and sufficiently close to it. How close to T_c one has to go depends on the ratio of the magnetic coherence length to the depth of information of the electron scattering experiment. This in turn depends very much on the system studied. The assumption of a localized Heisenberg model for a prototype itinerant ferromagnet like Ni appears questionable, and the numbers should be taken with due care. It is justified, for the time being, by the absence of realistic first-principles calculations of the magnetization profiles for semi-infinite 3d ferromagnets at arbitrary finite temperatures (see also the more sophisticated model calculations by Binder and Hohenberg /1974/).

In principle the magnetization profiles may be determined experimentally by the same trial-and-error procedure as used in structure determination by (spin-polarized) LEED: Theoretical results obtained for some assumed magnetization profile at some temperature are compared to experimental results obtained at the same temperature and the profile is varied until an optimum fit is achieved. It must not be overlooked, however, that in an actual experiment the magnetization profile is just one parameter among others that may change in a temperature dependent way. When we make a scattering experiment at fixed scattering geometry ($\underline{k},\underline{k}'$) and energy (E) and observe the exchange asymmetry as a function of temperature $A_{ex}(\underline{k},\underline{k}',E)$ = f(T) we would like to associate the function f(T) with the surface magnetization $M_1(T)$. This is, however, not allowed in the presence of strong multiple scattering because the magnetization profile changes as a function of temperature. How strong these effects may be, again depends on the ratio of the depth of information to the magnetic coherence length. This is, however, not the only reason why we should expect a non-linear behaviour. We know from the experience with non-magnetic solids /Kirschner and Feder, 1981/, that the temperature dependence of the multiple scattering causes spin polarization and asymmetry features to change in size and shape (cf. Fig. 4.8). Similar effects have to be expected in scattering from magnetic solids. Exchange asymmetry variations of multiple scattering origin will interfere with those of magnetic origin. A further effect may come from the thermal lattice expansion. Depending on the Curie temperature, the lattice may linearly expand by a percent or so over a full temperature scan. As the spin polarization is sensitive to structural parameters such as the lattice constant one must expect polarization features to shift on the energy scale, simultaneously with possible shape changes. Also, the lattice expansion may be different for the top layer(s). This causes additional structure-related changes of the exchange asymmetry. The top layer expansion may also be coupled with a variation of the exchange constant, leading to an additional decrease or increase of the surface magnetization. In view of all these possible complications one should not necessarily expect to measure the surface magnetization in an exchange asymmetry versus temperature run.

A demonstration for the above discussion may be found in Fig. 5.6, showing exchange asymmetry curves versus temperature for the specular beam from Fe(110) at various energies /Kirschner, 1984b/. The labels A, B ... refer to the energetic positions in Fig. 5.4a. The depth of information at the present geometry ranges from 1 to 2 layers around 80 eV to more than 5 layers below 10 eV. If nothing else happened one would therefore expect to measure essentially the surface magnetiza-

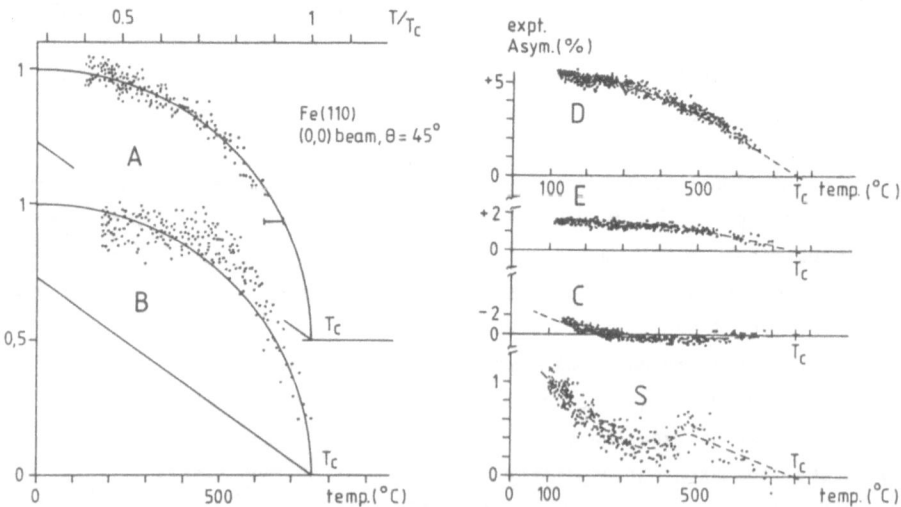

Fig. 5.6. Exchange asymmetry as a function of temperature for the specular beam from Fe(110). The labels A, B, ... refer to the energetic positions in the I-V and A-V spectra of Fig. 5.4a). Note the approximately bulk-like behaviour at A and B and the non-monotonic shape for C and S (for the conditions in S see text)

tion at point E and the bulk magnetization at points A and B. For the latter case the experimental points have been scaled to lie approximately on the bulk magnetization curve at 0.4 T_c. For comparison a linear dependence is indicated. Apparently, at points A and B the measured asymmetries fit the bulk magnetization curve quite well. The asymmetry versus temperature curves at points D and E clearly do not follow the bulk curve so that no attempt to fit them has been made. Visually, they appear to lie somewhere between the 'bulk' and the 'linear' behaviour, where the latter may be roughly taken to represent the surface magnetization (Binder and Hohenberg /1974/, Takeda and Fukuyama /1976/). Quite different behaviour is found at point C: Here the asymmetry starts from a negative value at room temperature, crosses zero far below T_c, reaches a positive maximum, and then gradually decays to zero again. Though point C is only about 3 eV from point D, the corresponding A_{ex} versus T curve clearly is not representative either of bulk or of surface magnetization. To demonstrate that this 'abnormal' behaviour is not a singular event, curve S shows the asymmetry for the surface covered with sulfur to saturation by segregation from the bulk. The non-monotonic behaviour is not due to sulfur segregation or dissolution during heating, but a structural transformation at the surface cannot be excluded. Though this curve has been taken at only 9 eV kinetic energy, its behaviour is clearly not 'bulk-like'. At the time being it cannot be said with any degree of reliability which of the above mentioned effects contributes how much to the observed behaviour. In order to extract the desired temperature dependence of the magnetization profile it will be necessary to measure large portions of asymmetry and intensity curves for a variety of temperatures. It should be emphasized that the above complications are inherent in the physics of the electron-surface interaction. They can in principle not be avoided by any experimental technique.

Still one may hope that the temperature effects of non-magnetic origin may be sufficiently constant over a small temperature range around the Curie temperature. Sufficiently close to T_c the magnetization profile may also be sufficiently constant and close to the surface magnetization. Though it is hard to decide on experimental grounds what is "sufficiently close to T_c", one may nevertheless do the experiment and see what happens. This was done by Alvarado et al. /1982b/ on Ni(001) and Ni(110). Fig. 5.7 shows results of measured exchange asymmetry versus reduced temperature $(1-T/T_c)$ on a log-log plot. From the slope of the straight line a critical exponent β_1 in the relation $M_1(t) \sim (1-T/T_c)^{\beta_1}$ may be determined. Within the experimental accuracy β_1 is around 0.8 for both surfaces. This value lies somewhere in between the critical exponents predicted for the Ising model (0.77 to 0.80) and for the Heisenberg model (0.81 to 0.88) (Reeve and Guttmann /1980/, Diehl and Dietrich /1981/, Guttmann et al. /1980/). It will have to be seen whether the experimental accuracy can be sufficiently improved to distinguish unequivocally between these predictions. It also remains to be tested whether the assumptions about the extension of the critical regime and the linearity of the

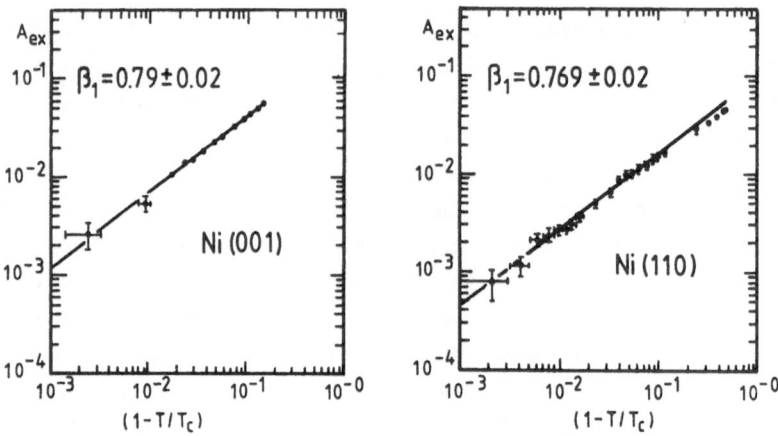

Fig. 5.7. Log-log plots of the exchange scattering asymmetry of the specular beam versus reduced temperature for Ni(001) and Ni(110)

exchange asymmetry underlying the experiment are valid up to $(1-T/T_c) \sim 0.1$ /Diehl and Eisenriegler, 1982/.

As mentioned several times it is the strong multiple scattering of electrons that makes the spin polarization sensitive to structural parameters. This fact, on the one hand, complicates the analysis by the need for adequate theoretical calculations. On the other hand it provides a means to analyze interrelationships between magnetic and structural quantities, such as a magnetization profile. If a kinematic approximation is applicable, much of the information is lost, but the analysis becomes simpler. If long range order is absent, multiple scattering is still strong but because of strong disorder the measured intensity is determined by the sum of intensities, rather than the sum of amplitudes. For a solid with homogeneous composition and magnetization the measured exchange asymmetry then is a measure of the atomic scattering factor for a single ferromagnetic atom (more precisely, for the atom plus its surrounding short range order if it exists). Under these conditions the exchange asymmetry probes an average magnetization over the depth of information. It may be used to study the temperature dependent near-surface magnetization of amorphous ferromagnets, a class of materials that has attracted much attention in recent years.

The low temperature behaviour of the magnetization of crystalline solids is well described by spin-wave theory. Spin waves, i.e. collective transversal excitations, are of low energy and are thermally excited. When present in sufficient number they reduce the magnetic moment, averaged over space and time. Neglecting the volume expansion with temperature one finds in good approximation /Keffer, 1966/:

$$M(T)/M(0) = 1 - \beta \cdot T^{3/2} - \gamma \cdot T^{5/2} \quad . \tag{5.1}$$

The first term is the well-known $T^{3/2}$-law of Bloch, while the second one takes into account higher-order terms in the magnon dispersion relation. It has been shown by Krey /1978/ that the low-temperature magnetization of an amorphous ferromagnetic glass is given by a law of the same form. The coefficient γ of the $T^{5/2}$ term has recently been determined by Majumdar et al. /1983/ for a series of Fe (B,C) glasses. The expansion of spin-wave theory to the case of surfaces was made by Mills and Maradudin /1967/, Mills /1982/, Mazur and Mills /1984/. In the low temperature regime the surface magnetization was predicted to follow also a $T^{3/2}$ law (to first order) but with a different factor β, making the surface magnetization to decay about twice as fast as the bulk magnetization. This prediction was tested by Pierce et al. /1982/ with polarized electron scattering from the metallic glass $Ni_{40} Fe_{40} B_{20}$. A result is shown in Fig. 5.8, comparing the bulk magnetization (solid line) to the average surface magnetization as obtained from the magnitude of the exchange asymmetry (crosses). The bulk magnetization curve was found to follow well the $T^{3/2}$-law over the temperature range studied, and so was the average surface magnetization (dashed line). A closer comparison revealed the surface magnetization to decay about three times faster than the bulk magnetization, a result not yet explained theoretically. Evidence for the $T^{5/2}$ term in (5.1) was not found.

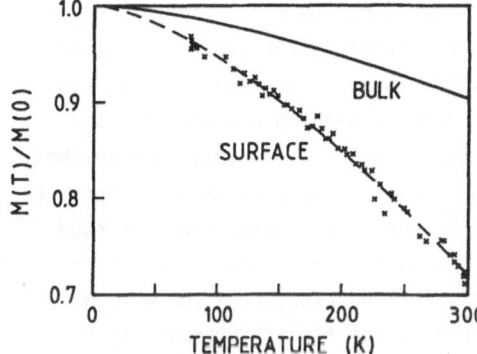

Fig. 5.8. Temperature dependence of relative bulk and surface magnetization in the ferromagnetic metallic glass $Fe_{40}Ni_{40}B_{20}$, as measured by elastic scattering of polarized electrons. The full ('bulk') and dashed lines both correspond to a $T^{3/2}$-law and the experimental data show that the surface magnetization decays with temperature according to the same law, but with a different prefactor

We saw in this section that the elastic scattering of polarized electrons or the polarization analysis of initially unpolarized electrons may serve as a probe for the magnetization in the near-surface region of solids. In crystals, multiple scattering makes the polarization structure-sensitive and complicates the analysis of temperature effects. On the other hand, the information content is larger than if a kinematic approximation is adequate (amorphous ferromagnets), and detailed magnetization profiles may be extracted via comparison to theoretical results of sufficient accuracy. In the next section we examine the elementary excitations in ferromagnets that are observable by inelastic scattering of polarized electrons.

5.1.2 Inelastic Scattering

We recall from Sect. 2.6 that the spectrum of inelastically diffracted electrons may approximately be decomposed into two parts: The one part contains electrons that were diffracted with their full primary energy and then lost energy in electronic excitations ('diffraction-before-loss'). The other part comprises electrons that lost energy first and then were diffracted with the reduced energy ('loss-before-diffraction'). Both processes are about equally probable (for not too large energy loss, as the diffraction cross section depends on the actual kinetic energy), and the loss spectrum will result from both processes by similar amounts. Sometimes - we discussed a particularly clear-cut example in Sect. 2.6.1 - the loss spectrum exhibits intensity structure that may clearly be attributed to loss-before-diffraction processes. Similar observations have also been reported for the spin-dependent intensity asymmetry in the energy loss spectrum from metallic glasses (Unguris et al. /1984/, Siegmann et al. /1981/). The asymmetry was to some extent similar to that of elastically scattered electrons of the same kinetic energy, though at reduced magnitude. The reasons for this behaviour are not yet quite clear, though it was argued that for large losses the energy dependence of the scattering cross section might for certain scattering conditions prefer the loss-first events in the inelastic spectrum.

The above results were obtained at rather limited energy resolution of the order of a few eV. The loss process energetically resolvable therefore was mainly plasmon excitation. We recall from Sect. 2.6 that the spin dependence of the inelastic mean free path, which is mainly determined by plasmon excitation, is rather small. The spin polarization effects therefore are essentially due to the elastic exchange scattering. In the following we will discuss an example, where at high energy resolution explicitly spin-dependent energy loss processes are studied. While in the low resolution experiments the asymmetry in the inelastic part is generally lower than in the elastic channel, at high resolution the asymmetry may be much larger at certain loss energies than in the elastic channel. In the present case the asymmetry is directly related to the spectrum of Stoner excitations at fixed momentum $q \approx 0$ of the electron-hole pairs. They were observed via exchange scattering contributions in the energy loss spectrum of spin-polarized electrons reflected specularly from a ferromagnetic Ni(110) surface /Kirschner et al., 1984/. The overall resolution in the experiment was 35 meV, obtained by mono-chromatizing the spin-polarized electrons emitted from a GaAsP photocathode. The monochromator system was the same as used for the differential energy- and spin-analysis of the GaAsP source described in Sect. 3.2. The scattered electrons were analyzed with respect to energy and intensity for opposite primary polarizations. A typical result is shown in Fig. 5.9a. We see that the loss intensity for down-spin electrons (i.e. oriented along the minority spin orientation) is larger than for up-spin electrons. The relative difference is largest around an energy loss of

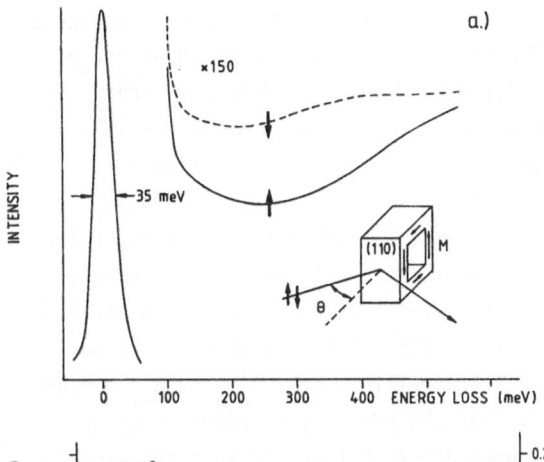

a.)

INTENSITY

×150

—35 meV—

(110) M

θ

0   100   200   300   400   ENERGY LOSS (meV)

Fig. 5.9. a) Typical electron energy
loss spectra for spin-up and spin-down
electrons scattered from Ni(110).
b) Intensity asymmetries as a function
of energy loss for reversed magnetizations at E_0 = 9.9 eV and Θ = 75°.
c) Asymmetry spectrum averaged over
different primary energies E_0. The
asymmetry peaks at ε = 0.28 eV with a
FWHM of 0.32 eV

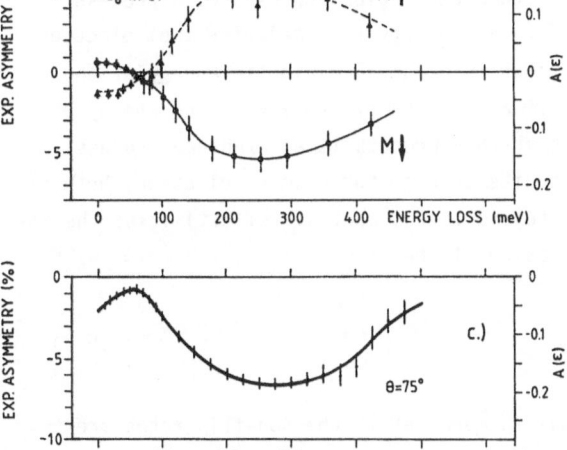

EXP. ASYMMETRY (%)

Θ=75°
E_0=9.9 eV

5-

0

-5-

M↑

M↓

b.)

0.2

0.1

0

-0.1

-0.2

A(ε)

0   100   200   300   400   ENERGY LOSS (meV)

EXP. ASYMMETRY (%)

0

-5-

-10-

Θ=75°

c.)

0

-0.1

-0.2

A(ε)

0   100   200   300   400   ENERGY LOSS (meV)

300 meV. The intensity difference is corrected for incomplete polarization of the
primary beam (~ 35 %). The intensity asymmetry in the loss spectrum, defined by
$A_{ex}(\varepsilon) = (I^{\uparrow}(\varepsilon) - I^{\downarrow}(\varepsilon)/(I^{\uparrow}(\varepsilon) + I^{\downarrow}(\varepsilon))$, is shown in Fig. 5.9b for a primary energy E_0 = 9.9 eV and an angle of incidence Θ = 75° for two orientations of the mag-
netization. The asymmetry is approximately symmetrical to the zero-line, though
not completely (note the crossing of the two curves around ε = 70 meV). In the
present scattering geometry the primary polarization vector is normal to the scattering plane, which makes the experiment sensitive to spin-orbit coupling. The
latter is independent of the magnetization and was measured to be rather small in
the present case. Therefore the data are not corrected. The asymmetry is seen to
be largest near ε ≈ 300 meV, larger than at ε = 0 and even of opposite sign. The
reversal of the curves with reversed magnetization proves the magnetic origin of
the asymmetry. The asymmetry below ε ≈ 50 meV is a remnant of the elastic exchange

scattering. It is due to multiple scattering and its sign and magnitude depends on
the energy and the angle of incidence. The large asymmetry feature, however, was
found to be nearly independent of the energy E_0 and angle Θ. Fig. 5.9c shows the
asymmetry averaged over a primary energy range from 5 eV to 15 eV at $\Theta = 75°$. The
shape of this curve (above $\sim$ 70 meV) represents approximately the spectrum of
Stoner excitations in Ni at zero momentum transfer. This is seen as follows: Among
the partial transition rates R (Fig. 2.14 in Sect. 2.6.2) there are three of little importance in Ni (a, c, e). They are those where in the final state a majority
electron resides above the Fermi level. These rates are small because in Ni, being
a 'saturated' ferromagnet, there are very few empty majority states available.
Among the remaining three rates there are two non-flip rates, $R_{nf}^{\uparrow}$ and $R_{nf}^{\downarrow}$ and one
flip rate $R_{f}^{\uparrow}$. We recall that 'flip' means that the emerging electron is of opposite spin as the incident electron. This does not mean that the incident electron
'really' flips its spin, like e.g. in magnon excitation. $R_{f}^{\downarrow}$ denotes a process
where an incident down-electron transfers its energy to a majority (up) electron
while occupying a state in the empty minority bands. The kinetic energy of the
ejected electron is smaller than the primary energy by the excitation energy of
the electron-hole pair created. Note that this process is an exchange process.
Though the emerging electron is of opposite spin to the primary electron, both of
the electrons preserve their spin. Expressed in the partial reflectivities the intensity asymmetry is given by an expression of the form:

$$ A = \frac{R_{nf}^{\uparrow} - R_{nf}^{\downarrow} - R_{f}^{\downarrow}}{R_{nf}^{\uparrow} + R_{nf}^{\downarrow} + R_{f}^{\downarrow}} \; . \qquad (5.2) $$

As discussed by Kirschner et al. /1984/ in more detail the non-flip rates are
nearly equal and contribute little to the measured asymmetry and we have approximately

$$ A_{ex}(\varepsilon) \approx -const \cdot R_{f}^{\downarrow} \; . \qquad (5.3) $$

The final state associated with this transition is characterized by a hole in an occupied majority band and an electron of opposite spin in an empty minority band.
This is exactly the configuration of a Stoner excitation and the experimental asymmetry will reflect the Stoner excitation spectrum. As the observations are made in
the specular beam, the Stoner excitation spectrum at $q \sim 0$ is monitored. Within a
model of rigidly shifted bands the Stoner density of states at $q = 0$ would be given
by a δ function at the energy of the exchange splitting Δ. If the exchange splitting
is $\underline{k}$-dependent near the Fermi surface, the δ function broadens into a distribution
which reflects the abundance distribution of the exchange splittings near the Fermi
surface. The experimental curve is, however, not directly proportional to the Stoner
density of states as the matrix elements may be energy and $\underline{k}$ dependent. The experimental distribution peaks around $\varepsilon = 300$ meV which represents the average exchange

splitting in Ni. This value agrees with most experimental data obtained by photo-emission at various locations in the Brillouin zone.

It may be noted that the above experiment was the first to observe the Stoner excitation spectrum at $q = 0$. This part of the ω,q space is not accessible to neutron scattering because of the large momentum of the neutron and the emission cha-racteristics of present day neutron sources. A complementary experiment was made by Hopster et al. /1984/ on an iron-based metallic glass, where unpolarized pri-mary electrons were used and the spin polarization was measured after energy loss. The expression for the polarization is of similar form as (5.2) and contains the same transition rates. In particular, if the polarization is essentially deter-mined by the flip rate $R_f^{\downarrow}$, the polarization should be positive. This was indeed observed experimentally. Spin-polarized electron energy loss spectroscopy thus ap-pears to be a suitable tool for the study of single-particle excitations in magne-tic materials. It will be interesting to see whether the energy resolution can be improved sufficiently to study collective excitations (magnons) at the surface.

5.1.3 Core Level Excitations

In this section we consider a special case of inelastic scattering, namely the ex-citation of core levels by electrons. This case differs from the previous one by the dispersionless character of the core levels, which justifies a 'random-k-ap-proximation'. This means that k-selection rules do not restrict the number of pos-sible excitations, and that transition rates are essentially determined by densi-ties of states or convolutions of them. These are not necessarily identical with ground-state densities of states as the creation of a hole may locally change the electronic system. Sometimes even 'atomic-like' features in cross-sections or lineshapes are observed. In non-magnetic materials all core levels are degenerate with respect to the spin. In ferromagnets this is not necessarily so as inner shells might be differently polarized by minority and majority electrons in the valence shell. A calculation by Krakauer et al. /1983/ predicts a splitting of the $3p_{3/2}$ level in Ni by roughly the same value as the average exchange splitting in the conduction band. This effect will probably be difficult to observe in Ni as the lifetime broadening tends to wash out such small differences. In iron, with its much larger exchange splitting (~ 2 eV), there is no evidence so far in con-ventional photoelectron spectroscopy for a splitting of this magnitude. Thus, if a splitting in the ground state exists it is most probably small enough to be ob-scured by lifetime broadening. Nevertheless, in this section we will find that the excitation of core levels by electrons is indeed spin-dependent in ferromagnets. In general terms, this phenomenon is due to the imbalance of the number of up-spin and down-spin electron states in ferromagnets and the resultant imbalance of dif-ferent excitation channels for primary electrons.

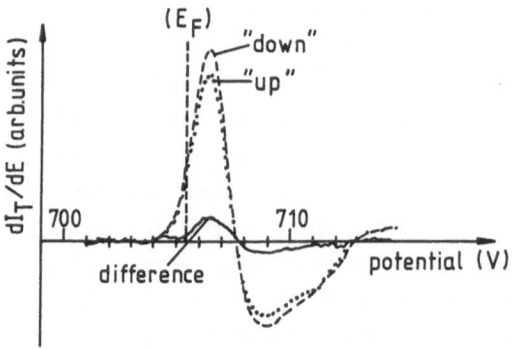

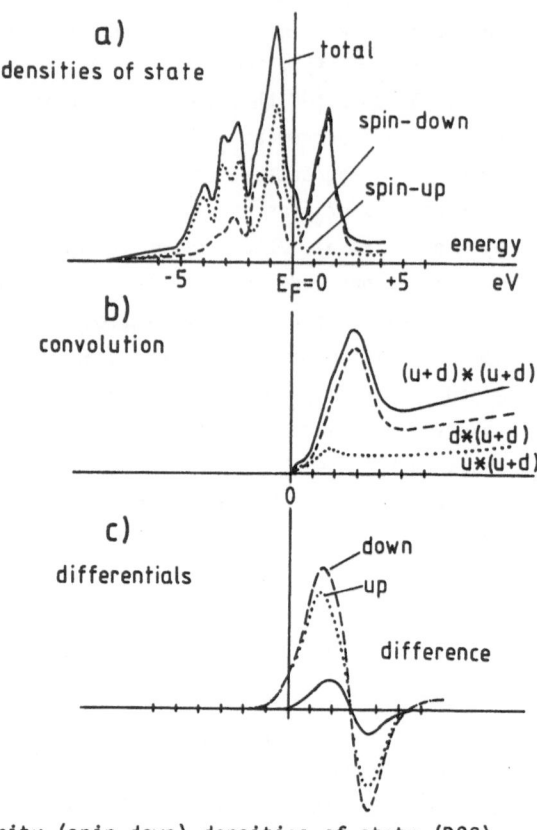

Fig. 5.10. Results of spin-polarized appearance potential spectroscopy of iron at the L_3 threshold. The vertical axis corresponds to the energy derivative of the net target current I_T for the primary electron spins oriented parallel to the minority states ("down") or to the majority states ("up") of Fe. The energetic position of the Fermi level has been obtained from Fig. 5.11c by matching the positive peaks

Fig. 5.11. a) Majority (spin-up) and minority (spin-down) densities of state (DOS) and the total DOS in Fe. b) Self-convolution of the empty total DOS (u+d)*(u+d) and corresponding convolutions of minority and majority DOS with the total DOS. c) Result of computer simulation of the experiment for opposite alignments of the primary electron spin. To be compared with Fig. 5.10

The first observation of this effect was made by Kirschner /1984c/ studying the threshold excitation of the Fe L_3 shell $(2p_{3/2})$ by spin-polarized primary electrons. The experimental set-up was identical to that shown in Fig. 5.16 below, with the photon detector not being used. Instead, the net current to the iron sample was monitored while the polarization of the primary electrons was reversed frequently and their kinetic energy was raised slowly. When the primary energy equals the binding energy of the L_3 shell, a slight step-like increase of the electron current leaving the crystal is observed, which is due to the emission of Auger electrons originating from the L_3 hole and to secondary electrons excited by them. The current increase is better visible when using electronic differentiation. A general discussion of this and related techniques may be found in a review by Kirschner /1977/. The experimental result for the two opposite spin orientations of the primary electrons is shown in Fig. 5.10. The observed signal is proportional to the excitation rate of the $2p_{3/2}$ level at threshold, and we see that it is larger for down-spin electrons (i.e. parallel to the minority electron spin) than for up-spin electrons. As an experimental proof for the dependence of this effect on the spin orientation of the target electrons, it was found that the difference curve changes sign when the magnetization is reversed. Also, the difference vanishes for normal incidence of the primary beam, as then spin polarization and magnetization are orthogonal. Qualitatively the same observations are made at the L_2 threshold. The signals are about half as large due to the smaller number of core electrons, and are somewhat broader due to the reduced lifetime of the L_2 level.

An explanation of these results was given by considering the spin-split densities of states (DOS) accessible to the two electrons involved in the transitions. Fig. 5.11a shows calculated spin-up and spin-down DOS for ferromagnetic iron (after Wang and Callaway /1977/). Excluding spin flips, a primary up-electron can only occupy a place in the up-DOS after transferring all its kinetic energy to a core electron. Within the approximation discussed above, for the core level electrons no such restriction is imposed. Assuming the excitation to be independent of the spin orientation of the core electrons (i.e. neglecting exchange processes), the available density of states for core electrons is given by the sum of empty spin-up DOS and spin-down DOS. The excitation rate will be proportional to the convolution of the empty up-DOS with the empty total DOS. This curve is shown in Fig. 5.11b, labelled u*(u+d). Similarly, for a down-spin primary electron the excitation rate will be proportional to the convolution of the empty down-DOS with the empty total DOS, labelled d*(u+d) in Fig. 5.11b. In the experiment, the primary beam is only partially polarized (P). It can be decomposed in a totally polarized fraction and a totally unpolarized fraction, with the ratio of P/(1-P). For the totally unpolarized fraction the excitation rate is given by the self-convolution of the empty total DOS, i.e. by the curve labelled (u+d)*(u+d) in Fig. 5.11b. In order to model the experiment properly, the convolution curves are added in proportion to the effective polarization. The result is convoluted by a Lorentzian curve with Γ = 0.25 eV to take the broadening due to the finite lifetime of the core level into account. It is further convoluted by a Gaussian curve with σ = 0.3 eV, which simulates the broadening due to the primary beam energy spread and due to the electronic differentiation. It was found that the final result for the spin-dependent effects is insensitive to the particular choice of these parameters within realistic limits. The 'up' and 'down' curves are differentiated with respect to energy, as is done in the experiment, and the final result is shown in Fig. 5.11c. When comparing the experimental and theoretical results (Fig. 5.10 and Fig. 5.11c), fairly good agreement is found, both for the lineshapes and the magnitude of the spin polarization effects. In the calculation, the ratio R_{th} between the maximum of the difference and the maximum of the sum is R_{th} = 0.095 compared to the experimental value of R_{exp} = 0.08 ± 0.01. It may be noted that even for a 100 % polarized primary beam the value $R_{th}^{100\,\%}$ ≈ 0.7 only would be obtained, due to the non-zero majority DOS above the Fermi level. The lineshape is asymmetric in theory and experiment. The full widths at half maximum of the positive excursions agree to within less than 0.1 eV. The negative excursion in the experimental curve is broader than in the theoretical one. This is attributed to the neglect of energy losses due to electron-hole and plasmon excitation in the model calculation. These become increasingly important with increasing distance from the Fermi level and tend to wash out the minimum in the convolution curves of Fig. 5.11b. We note that exchange processes need not be included in the model calculation to reproduce the experimental results. This does not mean that they are absent, rather the spin-split densities of states appear to play the dominant role at threshold.

When the primary energy is raised well above the threshold energy, the two electrons will no more reside near the Fermi level in the final state. Rather, the surplus energy may in principle be shared between them in arbitrary amounts. If there is a high density of states near E_F, however, one of them will preferentially populate this empty DOS. In an energy loss spectrum of the electrons leaving the sample there will be a discontinuity at an energy loss corresponding to the binding energy of the core level. In many cases the structure above the discontinuity (i.e. at larger energy losses) has been found to be closely related to the density of states above E_F (see Kirschner /1977/ and references therein). As this DOS is predominantly of minority type, it will be preferentially populated by minority electrons. Because of the possibility of exchange between the primary electron and the core electron, there exists a loss channel that prefers minority electrons from the primary beam to produce excitations. With polarized primary electrons one would observe stronger loss intensity for down-spin electrons, in analogy with the energy loss experiment discussed in Sect. 5.1.2. In fact, a level diagram like Fig. 2.15 would be appropriate also in this case if the lower energy levels were interpreted as core levels. In an alternative experiment one could use an unpolarized primary beam and analyze the electrons leaving the surface with respect to energy and spin. Because of the exchange process one should observe an enrichment in up-spin electrons at the energy loss corresponding to the binding energy of the core level. Such an experiment has recently been carried out by Mauri et al. /1984/ with the ferromagnetic metallic glass $Fe_{83}B_{17}$. At the ionization threshold of the Fe 3p level ($\sim$ 55 eV) they did indeed observe a positive polarization in the loss spectrum. A direct proof for the role of the exchange interaction was obtained by varying the primary energy from about 3 times to 50 times the threshold energy. The polarization gradually decreased with increasing energy, in qualitative agreement with the known energy dependence of the exchange interaction. The experiment was complicated by the interference of the direct excitation channel with an autoionization process, known as Fano resonance. Though a quantitative interpretation of these results is not yet at hand, they nicely complement the above threshold excitation experiment, demonstrating the spin-dependence of core level excitations in ferromagnets.

5.2 Electron Emission

In this section we will be concerned with the emission of spin-polarized electrons from ferromagnets and the information contained therein. As a prototype for core level emission we discuss Auger electrons from Fe which owe their spin polarization to the exchange coupling between the net spin of the 3d electrons and the inner core electrons. The spin polarization of secondary electrons holds promise for a new imaging technique of magnetic structures. The spin character and dispersion of valence bands have been probed by photoemission and their temperature behaviour has been studied.

5.2.1 Auger Electron Emission

In Sect. 5.1.2 we found that the excitation of core levels depends on the spin of the exciting electrons. We did not care for the filling of the core hole and the resultant electron or photon emission. Here we shall find that the Auger electrons are polarized, irrespective of the way in which the core hole has been generated. The first observation of this phenomenon has been made by Landolt and Mauri /1982/, theoretical treatments were given by Bennemann /1983/ and Kotani and Mizuta /1984/. The experiment was made on a $Fe_{83}B_{17}$ metallic glass, which makes the consideration of spin-dependent diffraction unnecessary. (Similar results have recently been found by the same experimental group on Ni(110) and Ni(001).) The sample was excited by unpolarized electrons of 2.9 keV energy at 70° angle of incidence and the Auger electrons were observed in normal emission. An experimental result is shown in Fig. 5.12 (after Landolt and Mauri /1982/).

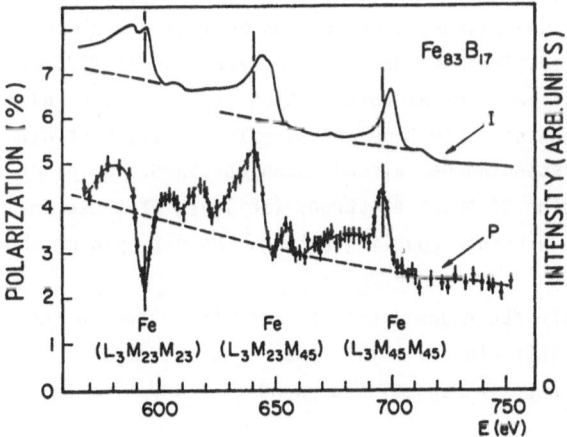

Fig. 5.12. Intensity I and spinpolarization P of Auger electrons from Fe in a $Fe_{83}B_{17}$ amorphous metal superimposed onto a smooth background of secondary electrons. The latter are assumed to contribute a polarization corresponding to the dashed line. Note the polarization in the $L_{23}M_{23}M_{23}$ groups of transitions, where the conduction band is not directly involved

The polarization features corresponding to the Auger lines ride on a slowly varying positive background which is due to inelastically scattered electrons. As the
relative intensity of the Auger lines is rather small (~ 20 % of the background intensity) their polarization is considerably reduced. When this background is corrected for, the polarizations are in the range of 15 - 20 %. First, the $L_3M_{23}M_{23}$
transition is discussed, which involves only transitions between core levels. The
final state in the Auger process is characterized by two holes in the M_{23}(3p) level,
which may be in the singlet or the triplet state. According to Hund's rule the Auger
electrons associated with the singlet state should have lower kinetic energy. The
low energy peak of the two-peak structure is therefore associated with a singlet final state, while the high energy peak is due to a triplet final state (which may
further split up due to exchange coupling with the 3d electrons). The initial 2p
hole may, in the case of a ferromagnet, couple to the net spin of the 3d electrons.
As shown by Kotani and Mizuta /1984/ the positive polarization of the singlet state
is due to the coupling of the 3d electrons to the 2p hole. The negative polarization
of the triplet peak mainly results from the coupling of the 3d electrons to the 3p
holes. The spin polarization therefore is predicted to vanish if the net 3d spin
goes to zero. The $L_3M_{45}M_{45}$ Auger line around 700 eV is the result of a final state
with two holes in the valence band. Like in Sect. 5.1.3 we conclude from the Coulomb
interaction being relatively small compared to the bandwidth, that the two holes are
nearly independent of each other. The lineshape and polarization are then essentially determined by convolutions of the occupied one-particle densities of states for
the two spin systems. A positive polarization is obtained, simply because there are
more majority spins, and a value of +26 % may be estimated for Fe, in reasonable
agreement with the experiment (+ 19 %). The $L_3M_{23}M_{45}$ line around 650 eV is characterized by a final state with one hole in the 3p shell and one in the valence band.
This case is intermediate between the two discussed before and an intermediate behaviour of the polarization is expected. For more details see Bennemann /1983/. At
present not all features in the rather complicated polarization structure are understood quantitatively (for the MMM lines see also Bader et al. /1983/, Zajak et al.
/1983/), but a fair qualitative understanding has already been obtained. The essential point is that the spin polarization of Auger electrons (at least of those involving core levels) results from the exchange coupling between the net spin of the
3d electrons and core electrons.

A point worth mentioning is that the KLL Auger electrons from the boron in the
sample were found to be unpolarized. This clearly illustrates that the Auger
process probes the local magnetic moments in magnetic compounds, in addition to
its element specifity.

5.2.2 Secondary Electron Emission

It has been observed in 1976 by Chrobok and Hofmann /1976/ that secondary electrons from a ferromagnet excited by unpolarized primary electrons are spin-polar-

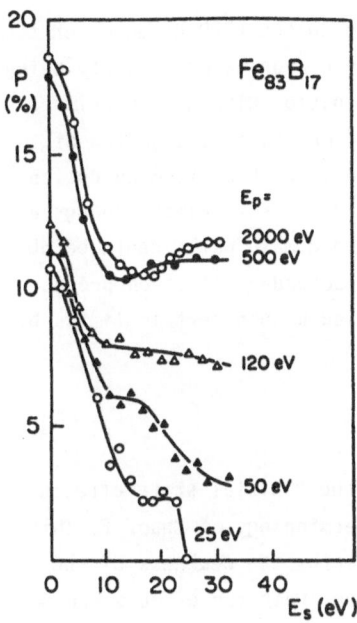

Fig. 5.13. Secondary electron spin polarization as a function of kinetic energy E_s, for various primary energies E_p. Note the increase of polarization towards small energies

ized. They observed polarization up to 32 % from EuO. This phenomenon has subsequently been investigated by several authors (Unguris et al. /1982a/, Kisker et al. /1982/, Hopster et al. /1982a/, Koike and Hayakawa /1983/) with similar results. As an example, Fig. 5.13 shows results for the ferromagnetic glass $Fe_{83}B_{17}$ /Landolt, 1984/. The spin polarization P of the secondaries is shown as a function of their energy E_s for various primary energies E_p from 25 eV to 2000 eV. First, we see that in all cases the maximum polarization is found for secondaries of near zero energy, reaching almost P = 20 % at high primary energy. This value is *higher* than expected from the average spin density. This phenomenon has been observed by a number of authors and several explanations have been put forward (Kisker et al. /1982/, Landolt /1984/). A full understanding has not yet been obtained, perhaps inelastic exchange scattering (see Sect. 5.1.2) may play an important role /Glazer and Tosatti, 1984/. The second effect we note in Fig. 5.13 is the dependence on the primary electron energy. This effect was attributed to the different probing depths at different energies. At low primary energy the penetration length of the primaries is rather shallow, it may even be smaller than the escape depth of secondary electrons. Therefore, the surface layer of the sample is probed, which is thought to have a reduced magnetization, due to chemical effects such as segregation or decomposition of the metallic glass at its surface. At primary energies of 500 eV and above the depth of information is determined by the escape depth of the secondaries, being of the order of 5 nm, and it is essentially the bulk magnetization that is probed.

The experimental data collected so far were mainly obtained by means of Mott detectors. This device appears not to be very suitable for a routine instrument due to its size and cost. Alternative approaches based on the LEED detector or its derivatives (Pierce and Unguris /1984/, private communication) are currently being developed /Kirschner, 1984d/. Such a detector in combination with a fine primary electron beam holds promise as a microanalytical tool for the investigation of surface magnetic properties at high resolution. Conventional beam forming optics for the exciting unpolarized beam could be used. At several keV primary energy a picture of the outer 5 nm or so of magnetic domains and domain walls could be obtained with a lateral resolution equal to that of the secondary electron production region (< 10 nm). The probing depth could be varied within certain limits by varying the primary beam energy.

5.2.3 Photoemission

In this section we consider experimental results for the "initial state effects" in momentum- and spin-resolved photoemission in our terminology of Chap. 2. This means that final state effects may be neglected, either by the weakness of the spin-orbit interaction or by working at normal emission. Operator effects are suppressed by using unpolarized or linearly polarized light. The spin polarization of photoelectrons then stems exclusively from the spin character of the occupied bands from where they are released. In this way the dispersion of occupied spin-split bands could in principle be mapped out in a way similar to that described below for spin-polarized inverse photoemission. This approach has not yet been persued very far, but there exist several measurements for Fe and Ni at particular points in the Brillouin zone. The polarization analysis is of great help in identifying transitions from particular bands without making recourse to theoretical bulk or surface band-structure calculations. Beyond this, the polarization analysis at elevated temperature yields important information about the evolution of exchange split bands when the Curie temperature is approached.

An example for momentum- and spin-resolved photoemission from Fe(001) is shown in Fig. 5.14 /Kisker et al., 1984/. The upper frame refers to photoemission at room temperature (T_C = 1043 K for bcc Fe). The data were obtained with linearly polarized synchrotron radiation in near-normal emission. At $h\nu$ = 60 eV transitions occur near the Γ point in the second Brillouin zone. In the experiment the intensity and spin polarization of electrons as a function of energy are measured. These two data sets are subsequently used to decompose the total intensity into its up-spin ($I^\uparrow$) and down-spin ($I_\downarrow$) contributions. The pair of exchange-split bands is designated by $\Gamma_{25}^{'\downarrow}$ (minority) and $\Gamma_{25}^{'\uparrow}$ (majority).

The transition $\Gamma_{12}^{\uparrow}$ should be forbidden, but may owe its existence to experimental imperfections (angular resolution, alignment, incomplete light polarization). Its minority counterpart lies above the Fermi level and cannot be observed in photoemission. The different widths of spin-up and spin-down peaks are explained by the reduced hole lifetime when going away from the Fermi level. The experimental ex-

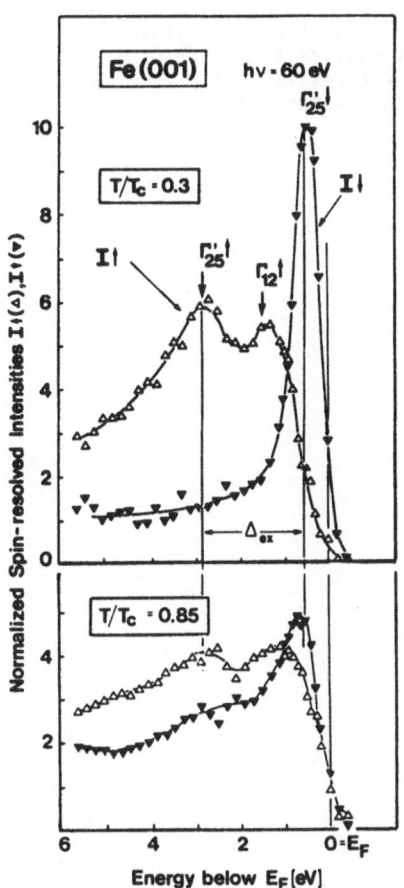

Fig. 5.14. Spin-resolved intensity curves, obtained from total intensity and spinpolarization as in Fig. 4.12 for Fe(001) with s-polarized light at reduced temperatures T/T_c = 0.3 and T/T_c = 0.85. Note that the exchange splitting Δ of the bands at Γ'_{25} is almost independent of temperature

change splitting as measured from the Γ'_{25} peak separation is 2.3 ± 0.2 eV at the Γ point. We wish to emphasize that such 'zero-temperature' results can well be described by the one-step theory of photoemission with a semi-empirical approximation of energy losses and their spin dependence /Feder et al., 1983b/. This holds true even for the relative intensities and the line-shapes of the peaks. The inelastic background was found to be dominated by majority electrons, with a polarization close to the average normalized difference of spin densities in the valence band. The lower panel in Fig. 5.14 shows data taken at elevated temperature, at T/T_c = 0.85. We first note the overall reduction of the intensities by roughly a a factor of two, which is due to the temperature dependent vibrations of the lattice, i.e. due to the structural disorder. Secondly, we find the spin-up and spin-down intensities to be much closer to each other than at low temperature. This is a consequence of the temperature dependent magnetic disorder, which reduces the average magnetization. As the spin polarization is measured with respect to an external axis of reference, given by the Mott detector in this case, the difference between spin-up and spin-down intensities necessarily vanishes when the macroscopic magnetization goes to zero, i.e. at the Curie temperature and above.

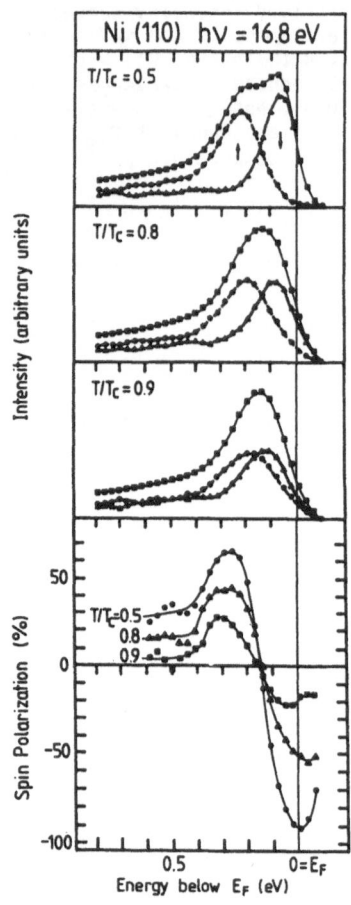

Intensity (arbitrary units)

Ni (110) hν = 16.8 eV

T/T_c = 0.5

T/T_c = 0.8

T/T_c = 0.9

Spin Polarization (%)

50

0

T/T_c = 0.5
0.8
0.9

-50

-100

0.5   0 = E_F

Energy below E_F (eV)

Fig. 5.15. Total intensities and spin-resolved photoemission intensities from Ni(110) at different reduced temperatures T/T_c. The lower panel shows the spinpolarization curves

We see that apart from these changes the spin-resolved intensity curves remain similar, notably that the energetic difference between the corresponding Γ'_{25} peaks changes only very little. This seems to indicate that the exchange splitting does not change very much at this location in the Brillouin zone and for this particular band (see also Sect. 5.3).

A second example for spin-polarized photoemission is shown in Fig. 5.15 for normal emission from Ni(110) /Hopster et al., 1983b/. Linearly polarized light at hν = 16.8 eV was used (from a Ne discharge lamp) and the geometry was chosen such that only bands with S_4 symmetry near the X point in Ni were excited. From top to bottom the figure shows total and spin-resolved partial intensities for three different temperatures and the polarization curves obtained at these temperatures. The exchange splitting at this point in the Brillouin zone was found to be 0.17 ± 0.01 eV at room temperature. This value is much smaller than in iron and the corresponding peaks in the total intensity are barely resolved, whereas in the spin-resolved intensities they are. With increasing temperature we note that the width of the total intensity curve decreases. This behaviour was observed pre-

viously in non spin-resolved spectra by Maetz et al. /1982/ and was tentatively
attributed to the growing contribution of an unpolarized central peak between the
two spin-split peaks. An analysis of this model in the light of the experimental
polarization data showed this probably to be too simple an explanation /Hopster et
al., 1983b/. One should keep in mind, however, that the very suggestive partial
intensity curves are no immediate proof against the existence of an unpolarized
contribution. They simply tell that the measured total intensity and polarization
curves would correctly result from their sum and normalized difference, respectively. A reduction of the polarization amplitudes could in principle also come
about by a third intensity curve with up and down spins in equal amounts. Under
conditions of limited resolution the decomposition therefore is not unique. We
shall come back to the discussion of the photoemission from Ni and in particular
from Fe when we know more about the empty bands and their temperature behaviour
(Sect. 5.3).

Momentum- and spin-resolved photoemission is a very valuable and sometimes even
indispensable tool for the detailed investigation of the electronic and magnetic
structure of ferromagnetic materials. We emphasize on this technique since it represents a significant advancement of photoelectron spectroscopy, but it does not
make yield spectroscopies obsolete. An important advantage of the latter is that
it can be carried out under conditions of a high external magnetic field. There
are numerous examples given in the reviews by Alvarado et al. /1978/ and Siegmann
et al. /1984/ where this feature is absolutely essential.

5.3 Isochromat Spectroscopy

Isochromat spectroscopy, also called inverse photoemission, essentially is the
spectroscopy of the short wavelength limit of the X-ray Bremsstrahlungsspectrum.
When performed at high energy, it provides information about the empty density of
states of solids, in a similar way as X-ray photoelectron spectroscopy does for
the occupied bands. When performed in the soft X-ray or far ultraviolet region
wave vector selection rules become increasingly important and close analogies to
momentum resolved photoemission appear (c.f. Sect. 2.5.2). In photoemission a photon is absorbed and an electron from the occupied bands is ejected with conservation of energy and quasi-momentum. In inverse photoemission an electron of definite momentum is injected into the solid and a photon is observed upon a transition of the electron into low energy electron states. Because of the small momentum of the photon the detection may be extended over the full half-space in front
of the sample without disturbing the momentum conservation in the electron transition process. The final state after photon emission may be chosen close to the
Fermi level. Therefore the important region between Fermi level and the vacuum
level can be studied by this technique, a region that is hardly accessible to

other techniques such as LEED, for example (see also Fig. 2.12). The technique in its non-polarized version has recently been reviewed by Scheidt /1983/ and Dose /1983/. As shown by Unguris et al. /1982b/ inverse photoemission may also be carried out with spin-polarized primary electrons. In ferromagnets the empty minority and majority bands can separately be mapped out in the Brillouin zone, as first demonstrated by Scheidt et al. /1983/ for iron.

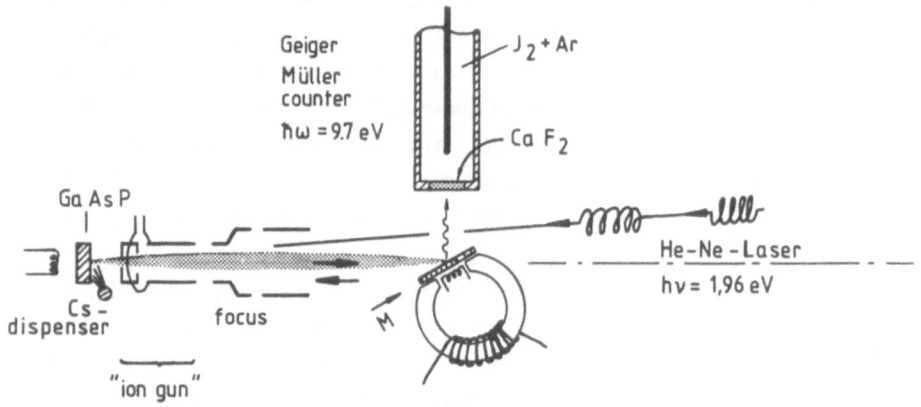

Fig. 5.16. Schematic drawing of a spin-polarized inverse photoemission experiment with a high intensity GaAsP polarized electron source and an energy-selective Geiger-Müller counter. The electron beam is polarized longitudinally

A schematic of the apparatus used for the latter experiment is shown in Fig. 5.16. A particularly simple GaAsP source of polarized electrons is used /Kirschner, 1984a/, in which the electron beam is not bent by 90°, in contrast to other sources described in the literature /Pierce et al., 1980/. Therefore the electron beam is longitudinally polarized and high intensities are obtained at low energy. With the present source more than 20 µA target current was obtained at 10 eV kinetic energy. High primary currents are required in inverse photoemission as the photon production is rather inefficient /Dose, 1983/. The overall intensity loss (number of incident electrons versus number of detected photons) is about ten orders of magnitude. If the plane of incidence coincides with a mirror plane of the crystal spin-orbit effects are suppressed. The iron single crystal forms part of a closed magnetic loop which eliminates stray fields effectively. The photons emitted by the sample are detected by an energy selective Geiger-Müller counter with a fixed pass energy of 9.7 ± 0.35 eV (FWHM). For more details of the detector see Scheidt /1983/. The simplicity of the apparatus is in remarkable contrast to the equipment used in spin-polarized photoemission from ferromagnets using a Mott detector /Raue et al., 1983/.

An example of the experimental results is reproduced from Scheidt et al. /1983/ in Fig. 5.17a and compared to recent theoretical results of Feder and Rodrigues /1984/ in Fig. 5.17b. For the calculations a non-relativistic one-step model of

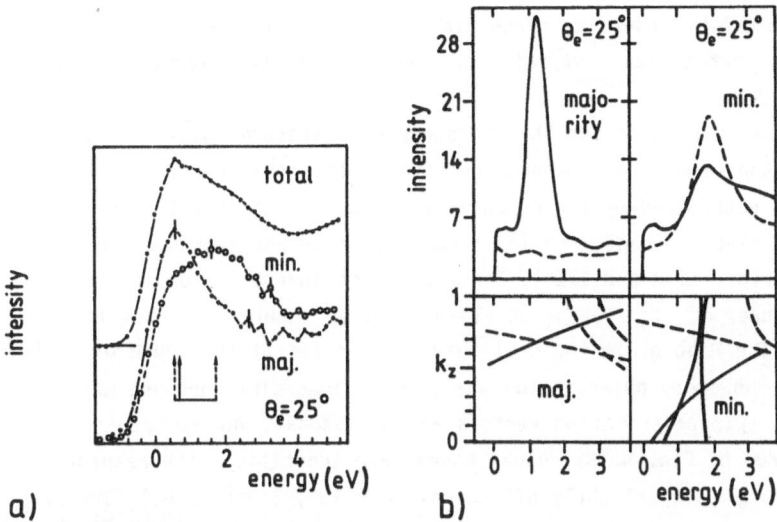

Fig. 5.17. Comparison of experimental a) and theoretical b), (upper panel) spin-resolved isochromat spectra from clean Fe(110). The full lines in the upper panel in b) refer to the p-polarized light intensity, the dashed lines to the s-polarized light. The lower panel in b) shows spin-resolved bulk bandstructures with the lower states (——) and the upper states shifted downwards in energy by the photon energy (---). In the theoretical intensity curves the inelastic background is not taken into account (after Scheidt et al. /1983/ and Feder and Rodrigues /1984/)

photoemission was adapted to isochromat spectroscopy. A non-relativistic approach appears adequate since polarization and magnetization both lie in the plane of incidence, which effectively suppresses interferences between spin-orbit and exchange interaction. Since the polarization properties of the emitted light are not analyzed in the experiment the s- and p-polarized intensities in the upper panel of Fig. 5.17b have to be summed up when comparing to the experiment. The lower panel in Fig. 5.17b shows the corresponding bulk band structure along the line $(k_\parallel, k_z)$ scanned in the isochromat energy run with $k_\parallel = [2(E+9.7-\phi)]^{1/2} \cdot \cos\theta_e$. The solid lines correspond to the low energy bands while the dashed lines represent the high energy bands shifted downwards in energy by $\hbar\omega = 9.7$ eV. The energy zero corresponds to the Fermi level. In the three-step-model of photoemission intensity peaks due to direct transitions are expected on the energy scale at the intersection points. This is seen to be veryfied to a large degree. Slight shifts of the peak positions are to be expected as the intensity spectra were calculated for lifetime-broadened states. While most features in the isochromat spectra can be ascribed to transitions between bulk energy bands, this is not always the case. For example, the hump near the Fermi level in the theoretical minority spectra has no counterpart in the bulk band structure. This feature may also be identified as a shoulder in the experimental minority intensity. As the calculated layer-resolved densities of states of minority character show a peak at this energy, it may be attributed to non-k-conserving transitions. By comparing bulk and surface densities of state a surface state of majority character has been identified by Feder and Rodrigues /1984/.

A question of major importance for the microscopic understanding of the magnetism in 3d metals is the temperature dependence of the electronic states. As the magnetism is largely determined by the holes in the d band, experimental investigations of the empty electron bands are least equally important as those for the occupied bands (cf. Figs. 5.14 and 5.15). The first such investigation has been carried out by Kirschner et al. /1984b/ on clean iron. An experimental result is shown in Fig. 5.18. The H point in the Brillouin zone is met at the angle of incidence $\Theta = 0°$, where, however, polarization analysis is impossible because magnetization and electron spin polarization vectors are orthogonal. The two peaks in the intensity spectrum of Fig. 5.18a, upper panel, are identified with respect to their spin character by going slightly off normal ($\Theta = 15°$, lower panel). The peak near the Fermi edge is of majority type, while the peak higher in energy is of minority type. The location in the Brillouin zone is indicated in the insert, and we find the exchange splitting of the two bands to be 1.6 ± 0.2 eV at room temper-

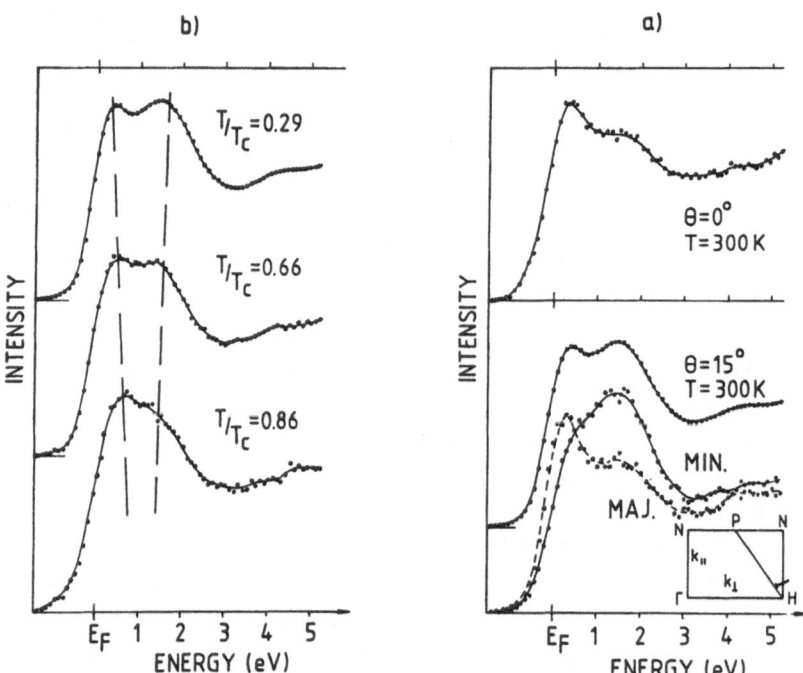

Fig. 5.18. Temperature dependent inverse photoemission data for Fe(100). a) Total intensity spectra at $\Theta = 0°$ and $\Theta = 15°$, taken at room temperature. The spin character of th two-peak structure is identified by the spin-resolved intensity spectra at $\Theta = 15°$. The insert shows the path in k space along which the isochromat spectra are measured. b) Intensity spectra at $\Theta = 15°$ for different reduced temperatures T/T_c. Note the merging of the two peaks into one when approaching T_c, indicating a collapsing empty band state near the H point of the Brillouin zone

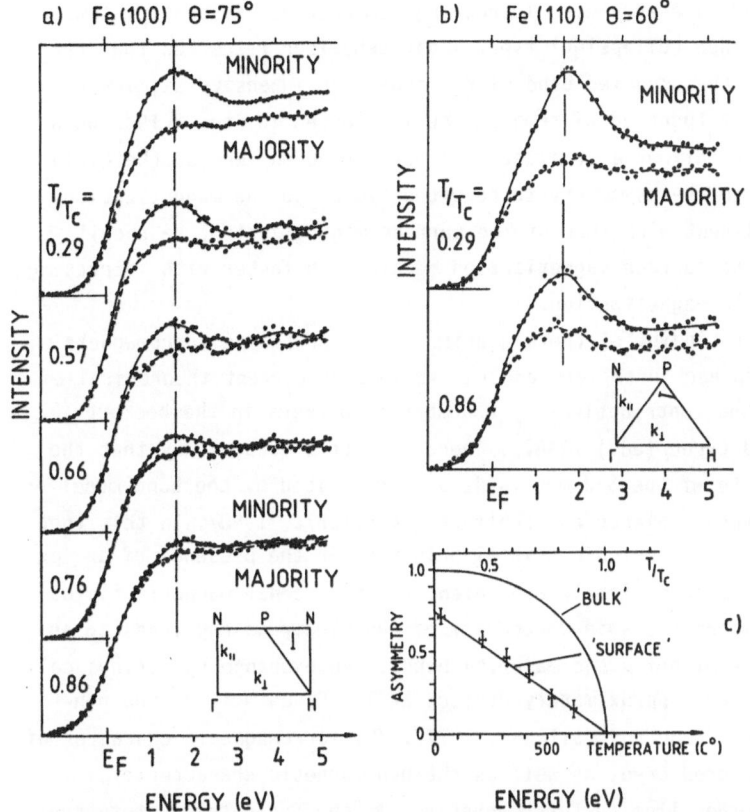

Fig. 5.19. a) Spin-resolved isochromat spectra from Fe(100) at $\Theta = 75°$ for various reduced temperatures. b) Same as in a) for the Fe(110) face at $\Theta = 60°$. Note that the maximum in the minority curves remains unchanged, indicating non-collapsing band states at the location in k space shown in the inserts. The asymmetry about 1.6 eV above E_F in a) is plotted as a function of temperature in c), indicating a surface magnetization behaviour

ature. The evolution of the spectra at $\Theta = 15°$ with increasing temperature is shown in Fig. 5.18b. There is a clear tendency of the two peaks to merge into one single peak on approaching the Curie temperature T_C, which indicates a 'collaps-ing' band state at this location in the Brillouin zone. As mentioned earlier, this behaviour would be expected from a Stoner model, where the exchange splitting goes to zero at T_C. Quite a different behaviour, however, is found in Fig. 5.19. The spectra in Fig. 5.19a,b refer to two different locations in k space (see inserts), sampled by using two different crystal surfaces. At these points, essentially a flat empty band of minority type is observed, while the corresponding majority band is below the Fermi level. We see that in both cases the minority band does not shift in energy but stays in place about 1.6 eV above E_F (Fig. 5.19a,b). The majority band remains below the Fermi level, otherwise a majority-type peak should appear at E_F and move up in energy with increasing temperature. Evidently, these band states are of the 'non-collapsing' type. Their behaviour resembles that observed in Fig. 5.14 for the occupied band of Γ'_{25} type. The intensity asymmetry in Fig. 5.19a at 1.6 eV as a function of temperature is plotted in Fig. 5.19c. When fitted at $T/T_c = 0.29$ it follows a straight line going through zero at the Curie temperature. If we assume the asymmetry to be proportional to the magnetization, this behaviour is consistent with that of the surface magnetization. We recall from Sect. 5.1.1 that the surface magnetization decays much faster with increasing temperature than the bulk magnetization.

The microscopic understanding of the magnetism of the prototype ferromagnets Fe, Co, and Ni is yet rather incomplete and the subject of current theoretical effort. See for example the contributions by a number of workers in the books by Moriya (ed.) /1981/ and Cyrot (ed.) /1982/. There is little doubt today that the ferromagnetic groundstate of the 3-d metals is well described by the band model or 'itinerant' model of Stoner, Slater and Wohlfarth /Wohlfahrt, 1980/. In this theory the d electrons move in a periodic crystal potential in the presence of an 'exchange field', which is mathematically equivalent to an internal magnetic field. This exchange field removes the spin degeneracy of the electrons and leads to an energetic splitting Δ of minority and majority bands. The exchange splittings calculated with this model are approximately correct at $T = 0$, as well as the non-integer magnetic moments /Wang and Callaway, 1977/. The ferromagnetic character of Fe, Co, Ni is correctly predicted, as well as the non-magnetic character of V or Pd, which are at the border line to ferromagnetism. At the Curie-temperature the exchange splitting Δ should vanish within this model, i.e. all exchange-split bands would collapse and there would not exist magnetic moments above T_c. The Curie temperature calculated with this model is of order $\Delta \sim k_B \cdot T_c$ and therefore far too high, even for Ni. In more recent theories, supported by neutron scattering experiments, the existence of local magnetic moments also above T_c is assumed. The loss of spontaneous magnetization above T_c may then be attributed to the disorder of the magnetic moments, a process that requires much less energy than destroying the magnetic moment itself. The Curie temperatures obtained from this approach are indeed much closer to reality, though not in perfect agreement. While there is common agreement about the existence of magnetic moments below and above T_c, there are controversial views about the nature of the magnetic structure. In the one approach termed 'local band theory' rather large domains with short range order around 10 Å are assumed, fluctuating in space and time. Within such domains a relatively well defined exchange-split band structure persists, similar to that in the ground state, with similar exchange splittings (Capellmann /1974/, Korenman et al. /1977a,b,c/, Ziegler /1982/). In the alternative view, the 'disordered local moments' model, the magnetic moments at different lattice sites are only weakly correlated (Edwards /1980/, Hubbard /1979/, Hasegawa /1979/, Heine et al. /1981/, Gyorffy et al. /1984/). With the disordered-local-moments model a specific prediction for the Bloch spectral function of Fe in the paramagnetic state has been made by Gyorffy et al. /1984/. A collapsing band state was predicted at the H point, in good qualitative agreement with the experimental result in Fig. 5.18. For the Γ'_{25} state a certain amount of exchange splitting was predicted to persist above T_c, which is in qualitative, though not quantitative, agreement with the photoemission measurements shown in Fig. 5.14. The persistence of spin-split peaks in photoemission was also predicted within the framework of the local band theory by Korenman and Prange /1980/, who pointed out that in between these peaks new structure may appear when approaching T_c. The shape changes mainly of the majority intensity in Fig. 5.19b may tentatively be ascribed to such effects, whereas the minority peak does not move. The photoemission results from Ni in Fig. 5.15 at first glance appear to be in contradiction with local band theory as the spin re-solved peaks clearly appear to merge. In a recent calculation for this case Koren-man and Prange /1984/ showed, however, that the experimental results could be well described by their theoretical approach. They also showed that for different values of $\underline{k}$ quite a different behaviour should be expected, for example even a four-peak structure. It appears remarkable that these quite different results were obtained with the assumption of a constant exchange splitting.

The complexity of this behaviour may be understood qualitatively in the follow-ing way. The magnetic fluctuations occur on a much longer time scale (of the order of spin wave frequencies) than the electronic transitions. Therefore, the eigen-states to be used when describing electronic processes are states computed for the instantaneous exchange field. These are not eigenstates of the momentum, which is fixed by the experiment. The experiment sees a combination of different eigen-states, averaged over a large number of exchange field configurations. This is equivalent to saying that in the presence of strong magnetic disorder $\underline{k}$ is no more a 'good' quantum number. The extent to which the spectral function of the bands is affected, depends on the band character in the vicinity of the $\underline{k}$ point chosen. For a flat band, such as investigated in Fig. 5.19 the mixing of energy eigenstates has little effect. If the band shows rapid dispersion, such as in the case of Ni at the X point and in Fe near the H point, the spectra are much stronger affected. Thus, some qualitative agreement between several experiments on the one hand, and both theoretical approaches on the other hand is reached. It will be a matter of comparison to detailed numerical calculations to decide which of the opposing theoretical views is the correct one, or whether they finally merge into a unified theory. Spin-polarized electron spectroscopies will certainly play a crucial role in this development.

6. Outlook

When extrapolating developments in a rapidly changing field such as the present one, any prediction is almost certainly bound to fail. Nonetheless, a personal view of probable or desirable future developments shall be given in the following.

On the experimental side, the search for more efficient sources and detectors will continue. The source intensities from the GaAs source and its derivatives are certainly quite satisfactory, but the degree of polarization should be improved. Though one knows well enough how total polarization should be obtained in principle (i.e. by removal of the band degeneracies near Γ), it is not easy to do in practice and several groups are working on this problem. For some special applications the field emission source will certainly be the first choice, but some development work is needed to transform it into a routine tool. Alternative sources based on photoemission from ferromagnets are worth being examined.

The LEED detector of polarized electrons has now been developed into a routine tool, but its development potential has not yet been explored. An improvement of the figure of merit by an order of magnitude seems to be within reach, but even then the detection of polarized electrons is a weak point in the whole field. This does provide a strong motivation to search for the 'ideal spin filter' which should operate in analogy to an optical light polarization analyzer. The operating principle of the field emission source shows that there do exist quite efficient spin filtering processes, but it is not known how one could exploit them in a polarization detector.

Even with the present equipment, however, there are quite a number of promising areas of research:

a) Elastic diffraction from non-magnetic materials could be exploited more extensively for surface structure analysis, in particular for systems incorporating heavy and light atoms. The spin dependence of threshold effects in LEED could yield a much better knowledge of the surface barrier. For these purposes the development of approximate but efficient computer codes appears desirable.

b) Spin-polarized photoemission with circularly polarized light is expected to become a major area of research. As demonstrated, it allows the experimental determination of band-structure symmetries and of hybridization between bands. The extension to more complicated band structures and two-dimensional systems is foreseeable. There are also exciting prospects for the characterization of structural disorder. As the development of a fully relativistic one-step theory of photoemission is underway, a deeper insight into the basic mechanisms of photoemission from solids is expected in the near future. When this technique is applied to ferromagnetic systems, a whole class of new phenomena may be anticipated. The exploitation of 'final state effects' might even replace an explicit spin polarization analysis by measuring relative intensity differences.

c) A major goal in the field of surface magnetism will be to determine the layer-resolved magnetization profile at arbitrary temperature. This has been demonstrated to be feasible in principle by polarized electron scattering, but the high precision required in theory and experiment calls for improved experimental and theoretical techniques. Alternative methods like positron diffraction, ion scattering with electron capture spectroscopy, or Penning ionization, when understood in more detail, will contribute much to this effort. The effects of chemisorbed or segregated foreign atoms on surface magnetism are still largely unknown.

d) The microscopic understanding of metallic magnetism in the bulk and at the surface is an area of research where spin-polarized new techniques will be indispensable. The evolution of spin-split bands with increasing temperature will further be studied with photoemission and inverse photoemission. The latter technique has the great advantage of being independent of a synchrotron light source.

e) Inelastic electron scattering is a very promising technique to study elementary excitations in ferromagnets, like the Stoner spectrum, and their temperature dependence. Perhaps surface magnons may be studied by the same technique. Work is underway towards the 'complete' experiment: Spin-polarized primary electrons plus spin analysis of the scattered electrons. Such an experiment will yield all flip and non-flip partial intensities and will also be applicable to non-magnetic samples. The theory of inelastic scattering of polarized electrons is beginning to emerge, and this technique may become as important for surfaces as neutron scattering has been for bulk studies.

f) There are also local probes of the magnetization at hand, in the form of Auger spectroscopy and appearance potential spectroscopy. When they are understood in more detail, they will allow the local magnetic behaviour to be analyzed in alloys undergoing compositional changes. Finally, the analysis of spin-polarized secondary electrons may provide a technique to image magnetic structures in real space at very high resolution.

References

Ackermann, B., R. Feder /1984/: Solid State Commun. 49, 489 (Sects. 2.1; 2.4.3)
Adams, D.L., U. Landman /1977/: Phys. Rev. B 15, 3775 (Sect. 4.1.2)
Adams, D.L., H.B. Nielsen, M.A. Van Hove /1979/: Phys. Rev. B 20, 4789 (Sect. 4.1.1)
Adams, D.L. /1979/: Surface Sci. 81, L337 (Sect. 4.1.2)
Adawi, I. /1964/: Phys. Rev. 134, A788 (Sect. 2.5.1)
Adnot, A., J.D. Carette /1977/: Phys. Rev. Lett. 38, 1084 (Sect. 4.1.3)
Alguard, M.J. et al. /1979/: Nucl. Instr. Meth. 163, 571 (Sect. 3.2)
Allenspach, R., F. Meier, D. Pescia /1983/: Phys. Rev. Lett. 51, 2148 (Sects. 2.5.2; 4.2.1)
Allenspach, R., F. Meier, D. Pescia /1984/: Appl. Phys. Lett. (in press) (Sect. 3.2.1)
Alvarado, S.F. /1979/: Z. Physik B 33, 51 (Sect. 2.5)
Alvarado, S.F., W. Eib, F. Meier, H.C. Siegmann, P. Zürcher /1978/: "Photoemission of Spinpolarized Electrons" in Photoemission and the Electronic Properties of Surfaces, ed. by B. Feuerbacher, B. Fitton, R.F. Willis (Wiley, New York) pp. 437-463 (Chap. 1; Sects. 2.5; 2.5.2; 5.2.3)
Alvarado, S.F. D. Weller /1984/: to be published (Sect. 5.1.1)
Alvarado, S.F. et al. /1981a/: Z. Physik B 44, 259 (Sect. 3.2.1)
Alvarado, S.F. et al. /1981b/: Z. Physik B 49, 129 (Sects. 2.4.2; 5.1.1)
Alvarado, S.F., F. Ciccacci, M. Campagna /1981/: Appl. Phys. Lett. 39, 615 (Sect. 3.2.1)
Alvarado, S.F. M. Campagna, F. Ciccacci, H. Hopster /1982/: J. Appl. Phys. 53, 7920 (Sect. 5.1.1)
Andersen, O.K. /1970/: Phys. Rev. B 2, 883 (Sect. 4.2.1)
Ashcroft, N.W. /1978/: "The Photoelectron Excitation Process" in Photoemission and the Electronic Properties of Surfaces, ed. by B. Feuerbacher, B. Fitton, R.F. Willis (Wiley, New York) (Sect. 2.5.1)

Bader, S.D., G. Zajak, J. Zak /1983/: Phys. Rev. Lett. 50, 1211 (Sect. 5.2.1)
Bartschat, K., G.F. Hanne, A. Wolcke, J. Kessler /1981/: Phys. Rev. Lett. 47, 997 (Sect. 2.6.1)
Bauer, P. /1980/: Report MPI f. Plasmaphysik, Garching, IPP9/34 (Sect. 4.1.1)
Bauer, P., R. Feder, N. Müller /1980/: Solid State Commun. 36, 249 (Sect. 4.1.1)
Bauer, P., W. Eckstein, N. Müller /1983/: Z. Physik B 52, 185 (Sect. 4.1.1)
Bell, R.L. /1973/: Negative Electron Affinity Devices (Clarendon Press, Oxford) (Sect. 3.2.1)
Bennemann, K.H. /1983/: Phys. Rev. B 28, 5304 (Sect. 5.2.1)
Berglund, C.N., W.E. Spicer /1964/: Phys. Rev. 136, A1030 and 1044 (Sect. 2.5.1)
Biefeld, R.M., P.L. Gourley, I.J. Fritz, G.C. Osbourn /1983/: Appl. Phys. Lett. 43, 759 (Sect. 3.2.1)
Binder, K., P.C. Hohenberg /1974/: Phys. Rev. B 9, 2194 (Sect. 5.1.1)
Binnig, G., H. Rohrer, Ch. Gerber, E. Weibel /1983/: Surface Sci. 131, L379 (Sect. 4.1.2)
Bir, G.L., A.G. Aronov, G.E. Pikus /1976/: Sov. Phys. JETP 42, 705 (Sect. 3.2.1)
Black, J.E., B. Laks, D.L. Mills /1980/: Phys. Rev. B 22, 1818 (Sect. 4.1.4)
Bohnen, K.P. /1981/: Primärbericht Kernforschunszentrum Karlsruhe Nr. 11/02/06 F (Sect. 4.1.2)

Borstel, G., M. Wöhlecke /1981/: Phys. Rev. B 24, 2321 (Sect. 2.5.2)
Borstel, G., M. Wöhlecke /1982/: Phys. Rev. B 26, 1148 (Sect. 2.5.2)
Borstel, G., M. Wöhlecke /1983/: Phys. Rev. B 28, 3153 (Sect. 2.5.2)
Bringer, A., M. Campagna, R. Feder, W. Gudat, E. Kisker, E. Kuhlmann /1979/: Phys.
    Rev. Lett. 42, 1705 (Sect. 2.6.2)
Bross, H. /1977/: Z. Physik B 28, 173 (Sect. 2.5.1)
Burke, P.G., J.F.B. Mitchell /1974/: J. Phys. B 7, 214 (Sect. 2.3)
Busch, G., M. Campagna, P. Cotti, H.C. Siegmann /1969/: Phys. Rev. Lett. 22, 597
    (Sect. 1, 2.5)

Callaway, J. /1964/: Energy Band Theory (Academic Press, New York) (Sect. 2.4.3)
Cardona, M., L. Ley (ed.) /1979/: "Photoemission in Solids" in Topics in Applied
    Physics, Vol 26, 27 (Springer, Berlin, Heidelberg, New York) (Sect. 2.5)
Calvert, R.L., G.J. Russel, D. Haneman /1977/: Phys. Rev. Lett. 39, 1226 (Sect.
    4.1.4)
Capellmann, H. /1974/: J. Phys. F 4, 1966 (Sect. 5.3)
Caroli, C., D. Lederer-Rozenblatt, B. Roulet, D. Saint James /1973/: Phys. Rev. B 8,
    4552 (Sect. 2.5.1)
Caroli, C., B. Roulet, D. Saint-James /1978/ in "Handbook of Surfaces and Inter-
    faces", chapter 3 ed. by L. Dobrzynski (Garland STPM Press, New York) (Sect.
    2.5.1)
Celotta, R.J., D.T. Pierce, G.-C. Wang, S.D. Bader, G.P. Felcher /1979/: Phys.
    Rev. Lett. 43, 728 (Chap. 1)
Celotta, R.J., D.T. Pierce /1980/: in Advances in Atomic and Molecular Phyiscs,
    Vol. 16, ed. by D.R. Bates, R. Bederson (Academic Press, New York) (Sect.
    1.3.2)
Celotta, R.J., D.T. Pierce, H.C. Siegmann, J. Unguris /1981/: Appl. Phys. Lett.
    38, 577 (Sect. 3.1.2)
Cherepkov, N.A. /1974/: Sov. Phys. JETP 38, 463 (Sect. 2.5)
Chrobok, G., M. Hofmann /1976/: Phys. Lett. 57A, 257 (Chap. 1; Sect. 5.2.2)
Clarke, L.J., L.M. de la Garza /1980/: Surface Sci. 99, 419 (Sect. 4.1.2)
Conrath, D. et al. /1979/: Appl. Phys. 20, 155 (Sect. 3.2.1)
Cyrot, M. (ed.) /1982/: Magnetism of Metals and Alloys (North-Holland, Amsterdam,
    New York) (Sect. 5.3)

Daniels, J., C.v. Festenberg, H. Raether, K. Zeppenfeld /1970/: Springer Tracts in
    Modern Physics, Vol. 54 (Springer, Berlin, Heidelberg, New York) (Sect. 2.6)
Darwin, C.G. /1928/: Proc. Roy. Soc. A 118, 654 (Sect. 2.2)
Davies, J.A. et al. /1978/: Surface Sci. 78, 274 (Sect. 4.1.1)
Davis, L.C., L.A. Feldkamp /1980/: Solid State Commun. 34, 141 (Sect. 2.5.1)
Davisson, C.J., L.H. Germer /1927/: Phys. Rev. 30, 705 (Chap. 1)
Davisson, C.J., L.H. Germer /1929/: Phys. Rev. 33, 760 (Chap. 1)
Davydow, A.S. /1967/: Quantenmechanik (VEB Deutscher Verlag der Wissenschaften,
    Berlin) (Sect. 2.3)
Debe, M.K., D.A. King /1979/: Surface Sci. 81, 193 (Sect. 4.1.2)
Dirac, P.A.M. /1928/: Proc. Roy. Soc. A 117, 610 (Chap. 1)
Diehl, H.W., S. Dietrich /1981/: Z. Physik B 42, 65 (Sect. 5.1.1)
Diehl, H.W., E. Eisenriegler /1982/: Phys. Rev. Lett. 48, 1767 (Sect. 5.1.1)
Dose, V. /1983/: Progr. in Surf. Sci. 13. 225 (Sects. 2.5.2; 4.1.3; 5.3)
Duane, W., F.L. Hunt /1915/: Phys. Rev. 6, 166 (Sect. 2.5.2)
Duke, C.B., G.E. Laramore /1971/: Phys. Rev. B 3, 3183 (Sect. 2.6.1)
Duke, C.B., U. Landman /1972/: Phys. Rev. B 6, 2968 (Sect. 2.6.1)
Dunlap, B.J. /1980/: Sol. State Commun. 35, 141 (Sect. 2.4.2)
D'yakonov, M.J., V.J. Perel' /1974/: Sov. Phys. Semicond. 7, 1551 (Sect. 3.2.1)

Eckstein, W. /1967/: Z. Physik 203, 59 (Chap. 1; Sect. 4.1.4)
Eckstein, W. /1970/: Report MPI f. Plasmaphysik, Garching IPP 7/1 (Sect. 2.4;
    Chap. 3; Sect. 4.1.4)
Edwards, D. /1980/: J. Magn. Magn. Mater. 15-18, 262 (Sect. 5.3)
Einstein, A. /1905/: Ann. Physik 17, 132 (Sect. 2.5)
Erbudak, M., N. Müller /1981/: Appl. Phys. Lett. 38, 575 (Sect. 3.1.2)
Erbudak, M., G. Ravano /1981/: J. Appl. Phys. 52, 5032 (Sect. 3.1.2)

Erbudak, M., G. Ravano, N. Müller /1982/: Phys. Lett. A 90, 62 (Sects. 3.1.1;
    3.2.2)
Erbudak, M., G. Ravano /1983/: Surface Sci. 126, 120 (Sect. 3.1.2)
Eyers, A. et al. /1983/: Nucl. Instr. Meth. 208, 303 (Sect. 4.2.1)
Eyers, A. et al. /1984/: Phys. Rev. Lett. 52, 1559 (Sect. 4.2.1)

Fan, H.Y. /1945/: Phys. Rev. 68, 43 (Sect. 2.5.1)
Fano, U. /1969/: Phys. Rev. 178, 131; Addendum 184, 250 (Sect. 2.5)
Farago, P.S. /1974/: J. Phys. B 7, 128 (Sect. 2.3)
Farrell, H.H., M.M. Traum, N.V. Smith, W.A. Roger, D.P. Woodruff, P.D. Johnson
    /1981/: Surface Sci. 102, 527 (Sect. 2.5.1)
Feder, R., H.J. Meister /1969/: Z. Physik 229, 309 (Sect. 2.4.3)
Feder, R. /1971/: phys. stat. sol. (b) 46, K31 (Chap. 1; Sect. 2.4.3)
Feder, R. /1972/: phys. stat. sol. (b) 49, 699 (Chap. 1; Sect. 2.4.3)
Feder, R. /1974/: phys. stat. sol. (b) 62, 135 (Sect. 4.1.3)
Feder, R., P.J. Jennings, R.O. Jones /1976/: Surface Sci. 61, 307 (Sect. 2.4.3)
Feder, R. /1977a/: Phys. Rev. B 15, 1751 (Sect. 4.1.3)
Feder, R. /1977b/: Solid State Commun. 21, 1091 (Sect. 4.2.2)
Feder, R., N. Müller, D. Wolf /1977/: Z. Physik B 28, 265 (Sect. 4.1.2)
Feder, R. /1979/: Solid State Commun. 31, 821 (Sect. 2.6.2)
Feder, R. /1980/: Phys. Lett. 78A, 103 (Sect. 2.4.2)
Feder, R., H. Pleyer, P. Bauer, N. Müller /1980/: Surface Sci. 109, 419 (Sect.
    4.1.1)
Feder, R. /1981/: J. Phys. C 14, 2049 (Sects. 2.3,4; 2.6.1,2)
Feder, R., J. Kirschner /1981a/: Surface Sci. 103, 75 (Sects. 2.4.2,3; 4.1.2)
Feder, R., J. Kirschner /1981b/: Solid State Commun. 40, 547 (Sects. 2.5.2; 4.2.2)
Feder, R., H. Pleyer /1982/: Surface Sci. 117, 285 (Sect. 5.1.1)
Feder, R., S.F. Alvarado, E. Tamura, E. Kisker /1983a/: Surface Sci. 127, 83
    (Sect. 5.1.1)
Feder, R. et al. /1983b/: Solid State Commun. 46, 619 (Sects. 2.5.2; 5.2.3)
Feder, R., F. Rosicky, B. Ackermann /1983c/: Z. Physik B 52, 31 and 53, 244 (Sect.
    2.4.3.)
Feder, R., A. Rodriguez /1984/: Solid State Commun. (in press) (Sect. 5.3)
Feibelmann, P.J., D.E. Eastman /1974/: Phys. Rev. B 10, 4932 (Sect. 2.5.1)
Feibelmann, P.J. /1975/: Phys. Rev. Lett. 34, 1092 (Sect. 2.5.1)
Feibelmann, P.J. /1976/: Phys. Rev. B 14, 762 (Sect. 2.5.1)
Feigerle, C.S., D.T. Pierce, A. Seiler, R.J. Celotta /1984/: Appl. Phys. Lett 44,
    866 (Sect. 3.2.1)
Felter, T.E., R.H. Barker, P.J. Estrup /1977/: Phys. Rev. Lett. 38, 1138 (Sect.
    4.1.2)
Feuchtwang, T.E., P.H. Cutler, D. Nagy /1978/: Surface Sci. 75, 490 (Sects. 2.5.2;
    4.2.1)
Feuerbacher, B., B. Fitton /1973/: Phys. Rev. Lett. 30, 923 (Sect. 2.5)
Feuerbacher, B., B. Fitton, R.F. Willis (eds.) /1978/: Photoemission and the Elec-
    tronic Properties of Surfaces (Wiley, New York) (Sect. 2.5)
Fishman, G., G. Lampel /1977/: Phys. Rev. B 16, 820 (Sect. 3.2.1)
Forstmann, F., H. Stenschke /1977/: Phys. Rev. Lett. 38, 1365 (Sect. 2.5.1)
Freeman, A.J. /1983/: J. Magn. Magnet. Mater. 35, 31 (Sect. 5.1.1)

Gadzuk, J.W. /1978/: Chapter 5 in Feuerbacher et al. /1978/ (Sect. 2.5.1)
Garwin, F., D.T. Pierce, H.C. Siegmann /1974/: Helv. Phys. Acta 47, 393 (Sect.
    3.2.1)
Gehrenbeck, R.K. /1973/: Ph.D. Thesis, University of Minnesota (Chap. 1)
Gerhardt, U., E. Dietz /1971/: Phys. Rev. Lett. 26, 1477 (Sect. 2.5)
Glazer, J., E. Tosatti /1984/: to be published (Sect. 5.2.2)
Goldmann, A. /1982/: Vakuum-Technik 31, 204 (Sect. 2.5.1)
Goudsmit, S.A., G.E. Uhlenbeck /1925/: Naturwissenschaften 13, 1953 (Chap. 1)
Gradmann, U. /1977/: J. Magn. Magnet. Mater. 6, 173 (Sect. 5.1.1)
Gradmann, U., G. Waller, R. Feder, E. Tamura /1983/: J. Magn. Magnet. Mater.
    31-34, 883 (Sect. 5.1.1)
Gray, L.G., M.W. Hart, F.B. Dunning, G.K. Walters /1984/: Rev. Sci. Instr. 55, 88
    (Sect. 3.1.1)

Grimes, C.C. /1978/: Surface Sci. 73, 379 (Sect. 4.1.3)
Grobman, W.D., D.E. Eastman, J.L. Freeouf /1975/: Phys. Rev. B 12, 4405 (Sect. 2.5.1)
Guttmann, A.J., G.M. Torrie, S.G. Whittington /1980/: J. Magn. Magn. Mater. 15-18, 1091 (Sect. 5.1.1)
Gyorffy, B.L. et al. /1984/: in The Electronic Structure of Complex Materials, ed. by P. Pliariseau, B.L. Gyorffy, W.N. Temmerman (ASI Series B), to be published (Sect. 5.3)

Hanne, G. /1976/: J. Phys. B 9, 805 (Sect. 2.6.1)
Hasegawa, H. /1980/: J. Phys. Soc. Japan 49, 178 and 963 (Sect. 5.3)
Hedvall, J., E. Hedin, O. Persson /1934/: Z. Physik. Chem. B 27, 196 (Sect. 5.1.1)
Heine, V., J.H. Samson, C.M.M. Nex /1981/: J. Phys. F 11, 2645 (Sect. 5.3)
Heinzmann, U., J. Kessler, J. Lorenz /1970/: Z. Physik 240, 42 (Sect. 2.5)
Heinzmann, U., K. Jost, J. Kessler, B. Ohnemus /1972/ Z. Physik 251, 354 (Sects. 2.5; 4.2.1)
Heinzmann, U., H. Heuer, J. Kessler /1975/: Phys. Rev. Lett. 34, 441 (Sect. 3.1)
Heinzmann, U., G. Schönhense, J. Kessler /1979/: Phys. Rev. Lett. 42, 1603 (Sect. 2.5)
Heinzmann, U., B. Osterheld, F. Schäfers /1982/: Nucl. Instr. Meth. 195, 395 (Sect. 4.2.1)
Helman, J.S., W. Baltensperger /1980/: Phys. Rev. B 22, 1300 (Sect. 2.6.1)
Herlt, G., R. Feder, G. Meister, E.G. Bauer /1981/: Solid State Commun. 38, 973 (Sect. 4.2.2)
Hertz, H. /1887/: Ann. Physik 31, 983 (Sect. 2.5)
Himpsel, F.J., W. Steinmann /1975/: Phys. Rev. Lett. 35, 1025 (Sect. 2.5.1)
Himpsel, F.J. /1983/: Advances in Physics 32, 1 (Sect. 2.5.1)
Hodge, L.A., T.J. Moravec, F.B. Dunning, G.K. Walters /1979/: Rev. Sci. Instr. 50, 5 (Sect. 3.1.1)
Hopster, H. et al. /1983a/: Phys. Rev. Lett. 50, 70 (Sect. 5.2.2)
Hopster, H. et al. /1983b/: Phys. Rev. Lett. 51, 829 (Sect. 5.2.3)
Hopster, H., R. Raue, R. Clauberg /1984/: Verhandl. DPG (VI) 19, 382 and Phys. Rev. Lett. 53, 695 (Sect. 5.1.2)
Hora, R. M. Scheffler /1984/: Phys. Rev. B 29, 692 (Sect. 2.5.1)
Hubbard, J. /1979/: Phys. Rev. B 19, 2626 and 20, 4584 (Sect. 5.3)
Humberg, P. /1981/: Ph.D. Thesis, Universität Münster, Germany (Sect. 4.2.1)
Hünlich, K. /1984/: Diploma Thesis, Rheinisch-Westfälische Technische Hochschule Aachen, Germany (Sect. 4.2.1)

Ibach, H., D.L. Mills /1982/: Electron Energy Loss Spectroscopy and Surface Vibrations (Academic Press, New York) (Sects. 2.6; 3.2.1)
Ichimura, S., M. Aratama, R. Shimizu /1980/: J. Appl. Phys. 51, 2853 (Sect. 2.2)
Ignatiev, A., J.B. Pendry, T.N. Rhodin /1971/: Phys. Rev. Lett. 26, 189 (Sect. 2.4)
Ignatiev, A., F. Jona, M. Debe, D.E. Johnson, S.J. White, D.P. Woodruff /1977/: J. Phys. C 10, 1109 (Sect. 4.1.2)

Jennings, P.J. /1970/: Surface Sci. 20, 18 (Chap. 1; Sect. 2.4.3)
Jennings, P.J. /1971/: Surface Sci. 27, 221 (Chap. 1; Sects. 2.4.3; 4.1.3)
Jennings, P.J., R.O. Jones /1978/: Surface Sci. 71, 101 (Sects. 2.4.3; 4.1.3)
Jepsen, D.W. /1979/: Phys. Rev. B 20, 402 (Sect. 2.5.1)
Jepsen, D.W. /1981/: Phys. Rev. B 22, 5701 (Sect. 2.4.3)
Jepsen, D.W., J. Madsen, O.K. Andersen /1982/: Phys. Rev. B 26, 2790 (Sect. 5.1.1)
Jonker, B.T., J.F. Morar, R.L. Park /1981/: Phys. Rev. B 24, 2951 (Sect. 4.1.3)
Jost, K., F. Kaussen, J. Kessler /1981/: J. Phys. E 14, 735 (Sect. 3.3.2)

Kalisvaart, M. et al. /1978/: Phys. Rev. B 17, 1570 (Sect. 4.1.1)
Kang, W.M., C.H. Li, S.Y. Tong /1980/: Solid State Commun. 36, 149 (Sect. 2.5.1)
Keffer, F. /1966/: in Handbuch der Physik, Vol. 18/2, ed. by S. Flügge, H.P.J. Wijn (Springer, Berlin, Heidelberg, New York) p. 1 (Sect. 5.1.1)
Keliher, P.J., R.E. Gleason, G.K. Walters /1975/: Phys. Rev. A 11, 1279 (Sect. 3.2)

Kenner, V.E., R.E. Allen /1973/: Phys. Rev. B 8, 2916 (Sect. 4.1.4)
Kesmodel, L.L., P.C. Stair, G.A. Somorjai /1977/: Surface Sci. 64, 342 (Sect.
    4.1.1)
Kessler, J. /1976/: Polarized Electrons (Springer, Berlin, Heidelberg, New York)
    (Chap. 1; Sects. 2.1,2,3,4; Chap. 3)
King, D.A., G. Thomas /1980/: Surface Sci. 92, 92 (Sect. 4.1.2)
Kirschner, J. /1977/: "Electron Excited Core Level Spectroscopies" in Electron
    Spectroscopy for Surface Analysis, ed. by H. Ibach, Topics in Current Physics,
    Vol. 4 (Springer Berlin, Heidelberg, New York) pp. 59-116 (Sect. 5.1.3)
Kirschner, J., R. Feder /1979/: Phys. Rev. Lett. 42, 1008 (Chap. 1; Sects. 2.4;
    3.1)
Kirschner, J., R. Feder /1981/: Surface Sci. 104, 448 (Sect. 4.1.4)
Kirschner, J., R. Feder, J.F. Wendelken /1981/: Phys. Rev. Lett. 47, 614 (Chap. 1;
    Sects. 4.2.1,2)
Kirschner, J. H.P. Oepen, H. Ibach /1983/: Appl. Phys. A 30, 177 (Sect. 3.2.1)
Kirschner, J. /1984a/: Surface Sci. 138, 191 (Sect. 5.1.1)
Kirschner, J. /1984b/: Phys. Rev. B 30, 415 (Sect. 5.1.1)
Kirschner, J. /1984c/: Solid State Commun. 49, 39 (Sect. 5.1.3)
Kirschner, J. /1984d/: Scanning Electron Microscopy '84 (SEM Inc. AMF O'Hare) (in
    press) (Sect. 5.2.2)
Kirschner, J., D. Rebenstorff, H. Ibach /1984/: Phys. Rev. Lett. 53, 698 (Sect.
    5.1.2)
Kirschner, J., M. Glöbl, V. Dose, H. Scheidt /1984/: Phys. Rev. Lett. 53, 612
    (Sect. 5.3)
Kisker, E. et al. /1978/: Phys. Rev. B 18, 2256 (Sect. 3.2.2)
Kisker, E. et al. /1980/: Phys. Rev. Lett. 45, 2053 (Sect. 3.2.2)
Kisker, E., W. Gudat, K. Schröder /1982/: Solid State Commun. 44, 623 (Sect.
    5.2.2)
Kisker, E., K. Schröder, M. Campagna, W. Gudat /1984/: to be published (Sect.
    5.2.3)
Kleinman, L. /1978/: Phys. Rev. B 17, 3666 (Sect. 2.6.1)
Koike, K., K. Hayakawa /1983/: Proceedings 9th Int. Vac. Congress and 5th Int.
    Conf. Solid Surfaces, Madrid, Spain, Extended Abstracts, p. 16 and Jap. J.
    Appl. Phys. 23, L85, L187 (Sect. 3.1.1)
Korenman, V., J.L. Murray, R.E. Prange /1977a,b,c/: Phys. Rev. B 16, (a) 4032,
    (b) 4048, (c) 4058 (Sect. 5.3)
Korenman, V., R.E. Prange /1980/: Phys. Rev. Lett. 44, 1291 (Sect. 5.3)
Korenman, V., R.E. Prange /1984/: to be published (Sect. 5.3)
Kotani, A., H. Mizuta /1984/: to be published (Sect. 5.2.1)
Koyama, K. /1975/: Z. Physik B 22, 337 (Sect. 2.5)
Koyama, K., H. Merz /1975/: Z. Physik B 20, 131 (Sect. 2.5)
Krakauer, H., A.J. Freeman, E. Wimmer /1983/: Phys. Rev. B 28, 610 (Sects.
    5.1.1,3)
Krey, U. /1978/: Z. Physik B 31, 247 (Sect. 5.1.1)
Kunz, C. (ed.) /1979/: Synchrotron Radiation, Topics in Current Physics, Vol. 10
    (Springer, Berlin, Heidelberg, New York) (Sect. 2.5)
Kuyatt, C.E. /1975/: Phys. Rev. B 12, 4581 (Chap. 1)

Lagally, M.G., T.C. Ngoc, M.B. Webb /1971/: Surface Sci. 25, 444 (Sects. 2.4.2;
    4.1.1)
Lagally, M.G. /1975/: "Surface Vibrations" in Surface Physics of Materials, ed.
    by J.M. Blakely (Academic Press, New York) (Sect. 4.1.4)
Lagally, M.G., J.A. Martin /1983/: Rev. Sci. Instr. 54, 1273 (Sect. 3.1)
Lampel, G, C. Weisbuch /1975/: Solid State Commun. 16, 877 (Sect. 3.2.1)
Landolt, M., M. Campagna /1977/: Phys. Rev. Lett. 39, 568 (Sect. 5.1.1)
Landolt, M., Y. Yafet, B Wilkens, M. Campagna /1978/: Solid State Commun. 25, 1141
    (Sect. 5.1.1)
Landolt, M., D. Mauri /1982/: Phys. Rev. Lett. 49, 1783 (Sect. 5.2.1)
Lang, J.K. et al. /1982/: Surface Sci. 123, 247 (Sect. 4.1.5)
Landolt, M. /1984/: to be published (Sect. 5.2.2)
Laramore, G.E., C.B. Duke /1971/: Phys. Rev. B 3, 3198 (Sect. 2.6.1)
Le Bossé, J.-C. /1981/: Thèse, Université Claude-Bernard-Lyon I, Lyon, France
    (Sect. 4.1.3)

Le Bossé, J.-C. et al. /1982/: J. Phys. C 15, 3425 (Sect. 4.1.3)
Lee, C.M. /1974/: Phys. Rev. A 10, 1598 (Sect. 2.5)
Levinson, H.J., E.W. Plummer, P.J. Feibelman /1979/: Phys. Rev. Lett. 43, 952 (Sect. 2.5.1)
Levinson, H.J., F. Greuter, E.W. Plummer /1983/: Phys. Rev. B 27, 727 (Sect. 2.5.1)
Liebermann, L.N., J. Clinton, D.M. Edwards, J. Mathon /1970/: Phys. Rev. Lett. 25, 232 (Sect. 5.1.1)
Liebsch, A. /1974/: Phys. Rev. Lett. 32, 1203 (Sect. 2.5.1)
Liebsch, A. /1976/: Phys. Rev. B 13, 544 (Sect. 2.5.2)
Liebsch, A. /1981/: Phys. Rev. B 23, 5203 (Sect. 2.5.1)
Loth, R. /1967/: Z. Physik 203, 66 (Chap. 1)
Lu, T.M., G.-C. Wang /1981/: Surface Sci. 107, 139 (Sect. 4.1.2)

MacDonald, A.H., J.M. Daams, S.H. Vosko, D.D. Koelling /1981/: Phys. Rev. B 23, 6377 (Sect. 4.2.1)
Maetz, C.J. et al. /1982/: Phys. Rev. Lett. 48, 1686 (Sect. 5.2.3)
Mahan, G.D. /1970/: Phys. Rev. B 2, 4334 (Sect. 2.5.1)
Mahan, A.H., T.W. Riddle, F.B. Dunning, G.K. Walters /1980/: Surface Sci. 93, 550 (Sect. 4.1.5)
Maison, D. /1966/: Phys. Lett. 19, 654 (Chap. 1)
Majumdar, A.K., U. Oestreich, D. Weschenfelder, F.E. Luborsky /1983/: Phys. Rev. B 27, 5618 (Sect. 5.1.1)
Malmström, G.J., Rundgren /1981/: J. Phys. C 14, 4937 (Sect. 4.1.3)
Marcus, P.M., J.E. Demuth, D.W. Jepsen /1975/: Surface Sci. 53, 501 (Sect. 4.1.2)
Marsh, F.S., D.A. King /1979a/: Surface Sci. 79, 445 (Sect. 4.1.2)
Marsh, F.S., D.A. King /1979b/: Surface Sci. 81, L343 (Sect. 4.1.2)
Marsh, F.S., M.K. Debe, D.A. King /1980/: J. Phys. C 13, 2799 (Sect. 4.1.2)
Matthew, J.A.D. /1982/: Phys. Rev. B 25, 3326 (Sect. 2.6.2)
Mauri, D., R. Allenspach, M. Landolt /1984/: Phys. Rev. Lett. 52, 152 (Sect. 5.1.3)
Mayer, H., H. Thomas /1957/: Z. Physik 147, 419 (Sect. 2.5.1)
Mazur, P., D.L. Mills /1984/: Phys. Rev. B 29, 5081 (Sect. 5.1.1)
McRae, E.G. /1979/: Rev. Mod. Phys. 51, 541 (Sects. 2.4.3; 4.1.3)
McRae, E.G., D.T. Pierce, G.-C. Wang, R.J. Celotta /1981/: Phys. Rev. B 24, 4230 (Sect. 4.1.3)
Meier, F., D. Pescia, T. Schriber /1982a/: Phys. Rev. Lett. 48, 645 (Sect. 5.1.1)
Meier, F., D. Pescia, M. Baumberger /1982b/: Phys. Rev. Lett. 49, 747 (Sect. 2.6)
Meier, F., D. Pescia /1984/: "Spinpolarized Photoemission by Optical Orientation" in: Optical Orientation ed. by F. Meier and B.P. Zakharchenya (North-Holland, Amsterdam) (Chap. 1; Sects. 2.5.2; 4.2.1)
Melmed, A.J., R.T. Tung, W.R. Graham, G.D.W. Smith /1979/: Phys. Rev. Lett. 43, 1521 (Sect. 4.1.2)
Merz, H., K. Ulmer /1966/: Z. Physik 197, 409 (Sect. 2.5.2)
Messiah, A. /1969/: Quantum Mechanics, chapter 15 (North-Holland, Amsterdam) (Sect. 2.4.2)
Mills, D.L., A.A. Maradudin /1967/: J. Phys. Chem. Solids 28, 1855 (Sect. 5.1.1)
Mills, D.L. /1982/: in Surface Excitations ed. by V.M. Agranovich, R. London (North-Holland, Amsterdam) (Sect. 2.6.2)
Mills, K.A. et al. /1980/: Phys. Rev. B 22, 581 (Sect. 4.2.1)
Moriya, T. (ed.) /1981/: Electron Correlation and Magnetism in Narrow Band Systems, Springer Series in Solid State Science, Vol. 29 (Springer, Berlin, Heidelberg, New York) (Sect. 5.3)
Mott, N.F. /1932/: Proc. Roy. Soc. 135, 429 (Chap. 1)
Müller, N. /1975/: Physics Lett. 54A, 415 (Sect. 5.1.1)
Müller, N., W. Eckstein, W. Heiland, W. Zinn /1972/: Phys. Rev. Lett. 29, 1651 (Chap. 1; Sect. 3.2.2)
Müller, N. /1979/: Report MPI für Plasmaphysik, Garching, IPP9/23 (Sects. 2.4; 4.1.2)
Müller, N., E. Erbudak, D. Wolf /1981/: Solid State Commun. 39, 1247 (Sect. 4.1.1)

Nilsson, P.O., N. Dahlbäck /1979/: Solid State Commun. $\underline{29}$, $\overline{303}$ (Sect. 2.5.1)
Nilsson, P.O., J. Kanski, C.G. Larson /1980/: Solid State Commun. $\underline{36}$, 111 (Sect. 2.5.1)
Noffke, J., L. Fritsche /1981/: J. Phys. C $\underline{14}$, 89 (Sect. 5.1.1)

Oepen, H.P. et al. /1983/: Proc. 9th Internat. Vacuum Congr. and 5th Int. Conf. Sol. Surfaces, Sept. 26-30, 1983, Madrid, Spain, Extended abstract SS.P.9.D (Sects. 3.1; 4.2.1)
Oepen, H.P. /1984/: Ph.D. Thesis, Rheinisch-Westfälische Technische Hochschule Aachen, Germany (Sects. 3.1; 4.2.1)
Ohnishi, S. A.J. Freeman, M. Weinert /1983/: Phys. Rev. B $\underline{28}$, 6741 (Sect. 5.1.1)
O'Neill, M.R., M. Kalisvaart, F.B. Dunning, G.K. Walters /1975/: Phys. Rev. Lett. $\underline{34}$, 1167 (Chap. 1; Sect. 3.1)

Palmberg, P.W., R.E. DeWames, L.A. Vredevoe /1968/: Phys. Rev. Lett. $\underline{21}$, 682 (Sect. 5.1)
Pendry, J.B. /1974/: Low Energy Electron Diffraction (Academic Press, London, New York) (Chap. 1; Sect. 4.1.4)
Pendry, J.B. /1976/: Surface Sci. $\underline{57}$, 679 (Sect. 2.5.1)
Persson, B.N.J. /1983/: Phys. Rev. Lett. $\underline{50}$, 1089 (Sect. 2.6)
Pescia, D., F. Meier /1982/: Surface Sci. $\underline{117}$, 302 (Sect. 4.2.1)
Pierce, D.T., F. Meier /1976/: Phys. Rev. B $\underline{13}$, 5484 (Sects. 2.5; 3.2.1)
Pierce, D.T. et al. /1980/: Rev. Sci. Instr. $\underline{51}$, 478 (Sects. 3.2.1; 5.3)
Pierce, D.T., S.M. Girvin, J. Unguris, R.J. Celotta /1981/: Rev. Sci. Instr. $\underline{52}$, 1437 (Sect. 3.1.2)
Pierce, D.T., R.J. Celotta /1982/: Advances in Electronics and Electron Physics, Vol. 56, ed. by C. Marton (Academic Press, New York) (Chap. 1; Sect. 3.2)
Pierce, D.T., R.J. Celotta, J. Unguris, H.C. Siegmann /1982/: Phys. Rev. B $\underline{26}$, 2566 (Sects. 2.6; 5.1.1)
Pines, D. /1963/: Elementary Excitations in Solids (Benjamin, New York) (Sects. 2.5; 2.5.1)
Plummer, E.W., W. Eberhardt /1982/: Advances Chem. Phys. $\underline{49}$, 533 (Sects. 2.5; 2.5.1)

Rahman, T., D.L. Mills /1980/: Phys. Rev. B $\underline{21}$, 1432 (Sect. 4.1.3)
Rasolt, M., H.L. Davis /1979/: Phys. Rev. B $\underline{20}$, 5059 (Sects. 2.4.3; 4.1.4)
Rasolt, M., H.L. Davis /1980/: Phys. Rev. B $\underline{21}$, 1445 (Sect. 4.1.4)
Rau, C., S. Eichner /1981/: Phys. Rev. Lett. $\underline{47}$, 939 (Chap. 3)
Rau, C. /1982/: J. Magn. Mag. Mat. 30, 141 (Chap. 3)
Raue, R., H. Hopster, R. Clauberg /1983/: Phys. Rev. Lett. $\underline{50}$, 1623 (Sect. 5.3)
Ravano, G., M. Erbudak, H.C. Siegmann /1982/: Phys. Rev. Lett. $\underline{49}$, 80, Erratum /1983/ Phys. Rev. Lett. 50 (Sects. 2.3; 3.1.2)
Read, M.N., G.J. Russell /1979/: Surface Sci. $\underline{88}$, 95 (Sect. 4.1.2)
Rebenstorff, D. /1984/: Report Jül-1896 (ISSN 0366-0885) KFA Jülich (Sect. 2.6.1)
Rebenstorff, D., J. Kirschner, H. Ibach /1984a/: to be published (Sect. 2.6.1)
Rebenstorff, D., J. Kirschner, H. Ibach /1984b/: Verh. DPG (VI) $\underline{19}$, 383 (Sects. 2.6.1; 5.1.1)
Reeve, J.S., A.J. Guttmann /1980/: Phys. Rev. Lett. $\underline{45}$, 1581 (Sect. 5.1.1)
Reichert, E., K. Zähringer /1982/: Appl. Phys. A $\underline{29}$, 191 (Sect. 3.2.1)
Reichertz, P.P., H.E. Farnsworth /1949/: Phys. Rev. 75, 1902 (Sect. 2.6.1)
Reihl, B., B.D. Dunlap /1980/: Appl. Phys. Lett. $\underline{37}$, 941 (Sect. 4.1.1)
Reihl, B. /1981/: Z. Physik B $\underline{41}$, 21 (Sect. 4.1.1)
Rendell, R.W., D.R. Penn /1980/: Phys. Rev. Lett. $\underline{45}$, 2057 (Sect. 2.6.2)
Riddle, T.W., A.H. Mahan, F.B. Dunning, G.K. Walters /1978/: J. Vac. Sci. Technol. 15, 1686 (Sect. 4.1.4)
Riddle, T.W., A.H. Mahan, F.B. Dunning, G.K. Walters /1979/: Surface Sci. $\underline{82}$, 517 (Sect. 4.1.5)
Ritchie, R.H., J.C. Ashley /1965/: J. Phys. Chem. Solids 26, 1689 (Sect. 2.6.2)
Robota, H., W. Vielhaber, G. Ertl /1984/: Surface Sci. $\underline{136}$, 111 (Sect. 5.1.1)
Rose, J.H., J.F. Dobson /1981/: Solid State Commun. $\underline{37}$, 91 (Sect. 4.1.2)
Rundgren, J., G. Malmström /1977a/: Phys. Rev. Lett. 38, 836 (Sect. 4.1.3)
Rundgren, J., G. Malmström /1977b/: J. Phys. C $\underline{10}$, 4671 (Sect. 4.1.3)

Sacchetti, F. /1980/: J. Phys. F 10, L231 (Sect. 2.5.1)
Saldana, X.I., J.S. Helman /1970/: Phys. Rev. B 16, 4978 (Sect. 2.6.2)
Schaich, W.L., N.W. Ashcroft /1971/: Phys. Rev. B 3, 2452 (Sect. 2.5.1)
Scheffler, M., K. Kambe, F. Forstmann /1978/: Solid State Commun. 25, 93 (Sect. 2.5.2)
Scheidt, H. /1983/: Fortschr. Phys. 31, 357 (Sects. 2.5.2; 5.3)
Scheidt, H., M. Glöbl, V. Dose, J. Kirschner /1983/: Phys. Rev. Lett. 51, 1688 (Sects. 2.5.2; 5.3)
Schiff, L. /1955/: Quantum Mechanics, chapters 10 and 14 (McGraw Hill, New York) (Sect. 2.5.1)
Schilling, J.S., M.B. Webb /1970/: Phys. Rev. B 2, 1665 (Sect. 2.4)
Schönhense, G. /1980/: Phys. Rev. Lett. 44, 640 (Sect. 2.5)
Seah, M.P., W.A. Dench /1979/: Surface Interface Analysis 1, 2 (Sect. 2.4.3)
Shull, C.G., C.T. Chase, F.E. Myers /1943/: Phys. Rev. 63, 29 (Chap. 1)
Siegmann, H.C., D.T. Pierce, R.J. Celotta /1981/: Phys. Rev. Lett. 46, 452 (Sects. 2.6.1,2; 3.1.2; 5.1.2)
Siegmann, H.C., F. Meier, M. Erbudak, M. Landolt /1984/: Advances in Electronics and Electron Physics (Academic Press, New York) (Chap. 2; Sects. 2.5.2; 5.1.1)
Smith, N.V., M.M. Traum /1973/: Phys. Rev. Lett. 31, 1247 (Sect. 2.5)
Smith, N.V., H.H. Farrell, M.M. Traum, D.P. Woodruff, D. Norman, M.S. Woolfson, B.W. Holland /1980/: Phys. Rev. B 21, 3119 (Sect. 2.5.2)
Spanjaard, D.J., D.W. Jepsen, P.M. Marcus /1977/: Phys. Rev. B 15, 1728 (Sect. 2.5.2)
Spicer, W.E. et al. /1979/: Surface Sci. 86, 763 (Sect. 3.2.1)
Spicer, W.E. et al. /1980/: Phys. Rev. Lett. 44, 420 (Sect. 3.2.1)
Stair, P.C. /1980/: Rev. Sci. Instr. 51, 132 (Sect. 3.1)
Steiner, P., H. Höchst, S. Hüfner /1979/: in Photoemission in Solids II, (chapter 7) M. Cardona and L. Ley, Topics in Applied Physics, Vol. 27 (Springer, Berlin, Heidelberg, New York) (Sect. 2.5.1)
Stocker, B.J. /1975/: Surface Sci. 47, 501 (Sect. 3.2.1)
Su, C.I., W.E. Spicer, J. Lindau /1983/: J. Appl. Phys. 54, 1413 (Sect. 3.2.1)
Sunjic, M., Z. Penzar /1984/: Solid State Commun. 49, 145 (Sect. 2.6.1)
Suzuki, T., N. Hirota, H. Tanaka, H. Watanabe /1971/: J. Phys. Soc. Japan 30, 888 (Sect. 5.1)

Takeda, T., H. Fukuyama /1976/: J. Phys. Soc. Japan 40, 925 (Sect. 5.1.1)
Tamura, E., R. Feder /1982/: Solid State Commun. 44, 1101 (Sect. 5.1.1)
Tamura, E., R. Feder /1984/: Surface Sci. 139, L191 (Sect. 5.1.1)
Tamura, E., B. Ackermann, R. Feder /1984/: J. Phys. C (in press) (Chap. 1)
Tong, S.Y., C.H. Li, D.L. Mills /1981/: Phys. Rev. B 24, 806 (Sect. 4.1.3)
Treglia, G., F. Ducastelle, D. Spanjaard /1980/: Phys. Rev. B 21, 3729 (Sect. 2.5.1)
Turnbull, J.C., H.E. Farnsworth /1938/: Phys. Rev. 54, 507 (Sect. 2.6.1)

Ulehla, M., H.L. Davis /1978/: J. Vac. Sci. Technol. 15, 642 (Sect. 4.1.4)
Unertl, W.N. /1978/: J. Vac. Sci. Technol. 15, 591 (Sect. 2.4)
Unguris, J., D.T. Pierce, A. Galejs, R.J. Celotta /1982a/: Phys. Rev. Lett. 49, 72 (Sect. 5.2.2)
Unguris, J. et al. /1982b/: Phys. Rev. Lett. 49, 1047 (Sects. 2.5.2; 5.3)
Unguris, J., D.T. Pierce, R.J. Celotta /1984/: Phys. Rev. B 29, 1381 (Sects. 2.6.1; 5.1.2)

Van Klinken, J. /1966/: Nucl. Phys. 75, 161 (Chap. 3)
Van der Veen, J.F., R.G. Smeenk, R.M. Tromp, F.W. Saris /1979/: Surface Sci. 79, 219 (Sect. 4.1.1)
Wainwright, P.F., M.J. Alguard, G. Baum, M.S. Lubell /1978/: Rev. Sci. Instr. 49, 571 (Sect. 5.1.1)
Walker, D.W. /1974/: J. Phys. B 7, L489 (Sect. 2.3)
Waller, G., U. Gradmann /1982/: Phys. Rev. B 26, 6330 (Sect. 5.1.1)
Wang, C.S., J. Callaway /1977/: Phys. Rev. B 15, 298 (Sects. 5.1.3; 5.3)
Wang, C.S., A.J. Freeman /1981/: Phys. Rev. B 24, 4364 (Sect. 5.1.1)

Wang, G.-C., R.J. Celotta, D.T. Pierce /1981/: Phys. Rev. B 23, 1761 (Sect. 3.1, 4.1.1)
Wang, G.-C., R.J. Celotta, D.T. Pierce /1982a/: Surface Sci. 119, 479 (Sect. 4.1.2)
Wang, G.-C., J. Unguris, D.T. Pierce /1982b/: Surface Sci. 114, L35 (Sect. 4.1.2,5)
Wang, S.-W. /1980/: Solid State Commun. 36, 847 (Sect. 5.1.1)
Webb, M.B., M.G. Lagally /1973/: Solid State Commun. 28, 301 (Sect. 4.1.4)
Weisskopf, V. /1935/: Z. Physik 93, 561 (Chap. 1)
Wendelken, J.F., J. Kirschner /1981/: Surface Sci. 110, 1 (Sect. 3.1, 4.1.5)
Westphal, D., A. Goldmann /1983/: Surface Sci. 126, 253 (Sect. 2.5.1)
Westphal, D., D. Spanjaard, A. Goldmann /1980/: J. Phys. C 13, 1361 (Sect. 2.5.1)
Willis, R.F., B. Feuerbacher, N.E. Christensen /1977/: Phys. Rev. Lett. 38, 1087 (Sect. 4.1.3)
Willis, R.F. /1981/: Phys. Rev. B 11, 909 (Sect. 4.1.3)
Wimmer, E., H. Krakauer, M. Weinert, A.J. Freeman /1981/: Phys. Rev. B 24, 864 (Sect. 5.1.1)
Wöhlecke, M., G. Borstel /1981a/: Phys. Rev. B 23, 980 (Sect. 2.5.2)
Wöhlecke, M., G. Borstel /1981b/: Phys. Rev. B 24, 2857 (Sect. 2.5.2)
Wöhlecke, M., G. Borstel /1981c/: phys. stat. sol. (b) 106, 593 (Sect. 2.5.2)
Wöhlecke, M., G. Borstel /1981d/: phys. stat. sol. (b) 107, 653 (Sect. 2.5.2)
Wöhlecke, M., G. Borstel /1984/: "Spinpolarized Photoelectrons and Crystal Symmetry", in Optical Orientation ed. by F. Meier, B.P. Zakharchenya (North-Holland, Amsterdam) (Sect. 2.5.2)
Wohlfahrt, E.P. /1980/: Physics of Transition Metals ed. by P. Rhodes (Inst. of Physics Conf. Ser. No. 55) p. 161 (Sect. 5.3)
Woodruff, D.P., B.W. Holland /1970/: Phys. Lett. 31A, 207 (Sect. 2.4.2, 4.1.1)

Yin, S., E. Tosatti /1981/: "Spin-flip inelastic scattering in electron energy loss spectroscopy of a ferromagnetic metal", Report IC/81/129, Internatl. Centre for Theoretical Physics, Miramare, Trieste (Sect. 2.6.2)

Zajak, G. J. Zak, S.D. Bader /1983/: Phys. Rev. Lett. 50, 1713 (Sect. 5.2.1)
Zanazzi, E., F. Jona /1977/: Surface Sci. 62, 61 (Sect. 4.1.2)
Ziegler, A. /1982/: Phys. Rev. Lett. 48, 695 (Sect. 5.3)
Zürcher, P., F. Meier /1979/: J. Appl. Phys. 50, 3687 (Sect. 3.2.1)
Zürcher, P., F. Meier, N.E. Christensen /1979/: Phys. Rev. Lett. 43, 54 (Sect. 4.2.1)

Subject Index

J. Kessler

Polarized Electrons

2nd edition. 1985. 157 figures. Approx. 300 pages.
(Springer Series on Atoms and Plasmas, Volume 1)
ISBN 3-540-15736-0

(The first edition was published in 1976 in "Texts and
Monographs in Physics", Springer-Verlag)

Contents:

Introduction. – Description of Polarized Electrons. –
Polarization Effects in Electron Scattering Caused by
Spin-Orbit Interaction. – Polarization Effects Caused
by Exchange Processes in Electron-Atom Scattering.
– Polarized Electrons by Ionization Processes. –
Further Relativistic Processes Involving Polarized
Electrons. – Polarized Electrons from Solids and
Surfaces. – Further Applications and Prospects. –
References. – Subject Index.

Electron Spectroscopy for Surface Analysis

Editor: **H. Ibach**

1977. 123 figures, 5 tables. XI, 255 pages. (Topics in
Current Physics, Volume 4)
ISBN 3-540-08078-3

Contents:

H. Ibach: Introduction. – *J. D. Carette, D. Roy:* Design
of Electron Spectrometers for Surface Analysis. –
J. Kirschner: Electron-Excited Core Level Spectroscopies. – *M. Henzler:* Electron Diffraction and Surface
Defect Structure. – *B. Feuerbacher, B. Fitton:* Photoemission Spectroscopy. – *H. Froitzheim:* Electron
Energy Loss Spectroscopy.

Springer-Verlag
Berlin
Heidelberg
New York
Tokyo

Springer Series in Surface Sciences

Editors: R. Gomer, G. Ertl

Springer-Verlag
Berlin
Heidelberg
New York
Tokyo